The Penguin Book of

THE NATURAL WORLD

WORLD

Penguin Books

Penguin Books Ltd., Harmondsworth, Middlesex, England
Penguin Books Inc., 7110 Ambassador Road, Baltimore,
Maryland 21207, U.S.A.
Penguin Books Australia Ltd., Ringwood, Victoria, Australia
Penguin Books Canada Ltd., 41 Steelcase Road West, Markham,
Ontario, Canada
Penguin Books (N.Z.) Ltd., 182–190 Wairau Road, Auckland 10,
New Zealand

First published 1976

Text set in Monophoto Times by Oliver Burridge Filmsetting Ltd.,
Crawley, Sussex, England
Printed by Jarrold & Sons Ltd., Norwich, England

Kestrel Books hardcover ISBN 0 7226 5203 8
Penguin Books paperback ISBN 0 14 00 3714 4

THE PENGUIN BOOK OF THE NATURAL WORLD

THE PENGUIN BOOK OF THE NATURAL WORLD

Editorial Consultant: Elizabeth Martin M.A. (Oxon)

Co-ordinating Editors: Sonya Larkin
Louise Bernbaum

Designer: Bridget Heal

Contents

Living things are all around us. Some of them, amoebas for example, are so small that they can only be seen through a high-powered microscope. Others, such as trees and whales, are far larger than man himself. Where these creatures came from, how they developed and their fascinating relationship to each other are described and explained in this book.

Young people today are exposed to an overwhelming spectrum of knowledge. They tend to acquire a patchwork of facts without an overall framework in which to set the diverse parts. *The Penguin Book of the Natural World* aims to provide a quick reference work which can be read from the beginning to trace the history of all living things, or just to look up a description of a biological fact.

The book is designed in logical sequence, identified by double-page spread numbers, with each spread forming a separate unit of information, complete with its illustration. Its vocabulary is straightforward and simple, and is accompanied by a detailed index to the whole work to increase its value as a reference book. The diagrams, charts and illustrations have all been specially chosen to explain the text and help the reader to clearly visualize where plants, animals and man fit into the natural world.

Introduction
1   our natural world

Why does a fish have gills? How does a plant breathe? What is blood made of? Biology, the science of living things, helps us find answers to questions like these. But what do we mean when we say that something is a living thing? For instance, we know that rabbits, earthworms and daisies are alive but that rocks and engines are not. To help them decide what is alive, biologists group together the features that living things have in common; they reproduce themselves, they take in food and oxygen from their surroundings, excrete their waste, have some kind of movement and are sensitive to their surroundings. Within the last 75 years, however, biologists have discovered the existence of minute organisms called viruses. Because viruses have only some of the characteristics of living organisms, it is not yet certain whether they should be called living organisms or merely very complex chemicals.

The many different kinds of plants and animals alive today have all developed from earlier ones by the process of evolution. This means that some organisms are related to each other because they have descended from a common ancestor. Plants and animals can be arranged in groups according to the features they have in common and this indicates how closely they are related. In this important branch of biology called taxonomy, or classification, groups of organisms are arranged in an order, beginning with the very simple and ending with the most complex, to form a kind of family tree.

Nearly 2000 years ago Aristotle, the Greek philosopher, tried to classify living organisms. He divided the living world into three main groups – plants, plant animals, which included the sponges and sea anemones, and animals. But in recent times a more complex system is used, based mainly on the work of the 18th century Swedish naturalist, Linnaeus. To illustrate how the classification system works, take as an example the cultivated garden tea rose called 'Waltztime'. All the plants in the plant kingdom are first divided into five large groups called divisions. Spermatophyta is the division which contains the seed-bearing plants – herbaceous plants, trees and shrubs. A rose is therefore a spermatophyte because it produces seeds. The divisions are split into subdivisions and subdivisions into classes. The subdivision Angiospermae contains all the flowering plants. It is divided into the classes Monocotyledonae (or monocots), having one seed leaf, and Dicotyledonae (or dicots), having two. The rose is classed as a dicot because it is a flowering plant with two seed leaves.

The rose belongs to the Rosales order, which includes many important fruit trees distributed all over the world. Orders are made up of a number of families. The rose is placed in the family Rosaceae whose members all have four or five petals and four or five sepals. Families, in turn, are broken down into genera (singular: genus). The rose belongs to the genus *Rosa* which all produce hips, a kind of fruit. The genera, in turn, are each divided into a number of species. The cultivated rose is a hybrid of the species *Rosa odorata* whose flowers all have a particular scent. The species is the only group of organisms which can breed with one another to produce healthy fertile offspring. Sometimes members of a genus can breed with one another but the offspring, called hybrids, are usually sterile.

Animal classification works in the same way as plant classification, except that the overall animal groups which are similar to divisions in plants are called phyla (singular: phylum).

Linnaeus invented a system by which every plant and animal is given two Latin names: the first indicates the genus and the second the species to which it belongs. Thus the domestic dog is *Canis familiaris*, a species of the genus *Canis*, which also includes the wolves. This system avoids confusion between the common names of an organism which may vary from one country to another.

The field of natural science is very wide, and therefore biology is subdivided into many branches. The three main divisions are botany, the study of plants; zoology, the study of animals, and human biology, the study of the human body and the relationships between man and animals. Subdivisions include cell biology, which explores the structure and function of the cells that make up the bodies of plants and animals; genetics, which explains why offspring resemble their parents, and ecology, the study of how plants and animals live together in their natural communities and how the activities of man affect them.

Introduction
2 putting things in order

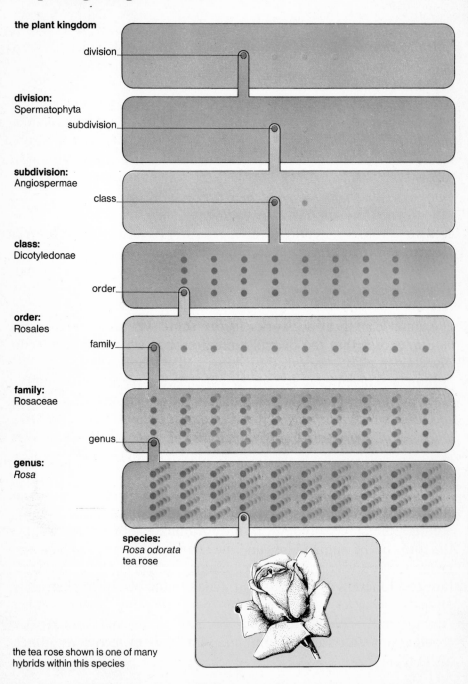

the plant kingdom

division

division:
Spermatophyta

subdivision

subdivision:
Angiospermae

class

class:
Dicotyledonae

order

order:
Rosales

family

family:
Rosaceae

genus

genus:
Rosa

species:
Rosa odorata
tea rose

the tea rose shown is one of many
hybrids within this species

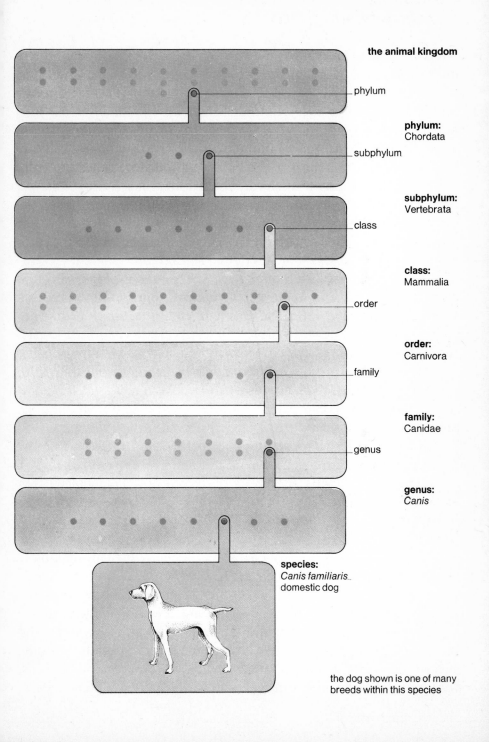

the animal kingdom

phylum:
Chordata

subphylum:
Vertebrata

class:
Mammalia

order:
Carnivora

family:
Canidae

genus:
Canis

species:
Canis familiaris
domestic dog

phylum

subphylum

class

order

family

genus

the dog shown is one of many
breeds within this species

Cells: the smallest units of life
3   their structure

All living things, except for the very smallest, are made up of many units called cells. Although they basically contain the same structures, cells are very varied in shape, size and function. There are many millions of cells in every human being, all too small to be seen individually without a microscope. Sperm cells are particularly small, about five-hundredths of a millimetre long. The very largest cells are bird's eggs, in which most of the bulk is a food store called the yolk.

Each cell contains a nucleus, which looks like a dot under an ordinary microscope and is embedded in a jelly-like mass called the cytoplasm. Using an electron microscope, which gives a greater magnification, various structures (organelles) can be seen in the cytoplasm. Each of these organelles does a special job.

The nucleus controls the shape, size and functions of the cell and it contains the hereditary material. Around the outside of the cell is a very thin membrane called the plasma membrane, which is made of protein and fat. It acts as a sieve, allowing certain chemicals to pass through, while at the same time keeping others out. Within the cell is a complex series of channels bounded by membranes, the endoplasmic reticulum, which acts as a sieve and as a means of transport for substances in the cytoplasm. Another membrane system, the golgi apparatus, secretes substances made by the cell.

Among the most important cell structures in the cytoplasm are mitochondria and ribosomes. The mitochondria are organelles in which the complicated processes of respiration take place; they release the energy that keeps the cell alive. The ribosomes are attached to the endoplasmic reticulum. Ribosomes, which are made of RNA (ribonucleic acid), are very small structures that help to make proteins – a process controlled by the nucleus. Protein-making is vital to the working of the cell because some proteins are enzymes, the catalysts for all the processes of life (a catalyst is a substance that speeds up a reaction without itself being used up). Nuclei, mitochondria, ribosomes and membranes are found in both animal and plant cells. Some cells contain other structures or materials for doing special jobs. For instance, haemoglobin is found in blood cells, contracting strands in muscle cells and chloroplasts in plant cells.

golgi apparatus   endoplasmic reticulum   nucleolus   nucleus   mitochondria

centriole

vesicle

microtubules   lysosomes   plasma membrane   polysomes   ribosomes

Cells: the smallest units of life
4 differences in plants

There are some structures in plant cells that are not found in animal cells. The easiest to see is the cell wall made of cellulose, which is deposited in layers around the outside of the outer cell membrane. The layer between two neighbouring cell walls is called the middle lamella and is made of calcium pectate. Cellulose is a tough but flexible material made up of long, cross-linked chains of glucose molecules. Many animals, including man, cannot digest cellulose when they eat plants. But even when the undigested cellulose passes through the body it serves a purpose: it provides bulk (roughage) in the intestines so that muscles do not become flabby through lack of use and cause constipation. Herbivores, such as rabbits and cattle, have special chambers in their digestive canals in which bacteria act upon the cellulose to make it digestible.

Some plant cell walls are further strengthened by another material called lignin which is laid on top of the cellulose cell wall. Water-conducting cells (xylem) are strengthened like this and form the wood that makes up the main part of a tree trunk. Most of the woody cells are dead and hollow and consist only of the wall.

A group of plant cells, enlarged to show their construction

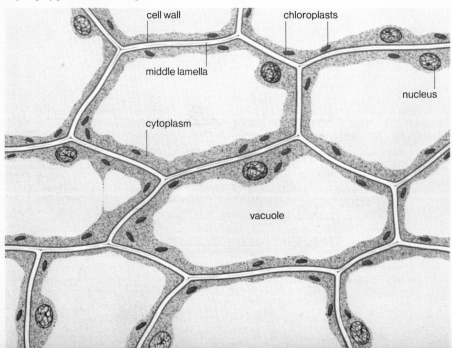

Much of a plant cell consists of a fluid-filled space, the vacuole, one of whose functions is to keep the cell 'inflated'. The vacuole contains a solution called cell sap. In some plants this may be coloured, such as the red sap in beetroots. The colours in some flowers, especially red and blue ones, are due to coloured cell sap in the cells of the petals. Many plant cells also contain chloroplasts. These are oval structures that contain chlorophyll, the green pigment which gives plants their green colour and is essential for photosynthesis. Chlorophyll traps the light energy necessary for making food to build more tissue. Chloroplasts can be seen clearly with the ordinary light microscope. There are other structures in some plant cells called chromoplasts, which are similar to chloroplasts but contain yellow and red colouring matter often found in the petals of flowers.

Plant cells have different shapes according to the jobs they do. Xylem and phloem cells are long because they form systems of tubes that transport substances through the plant. Some cells are specialized as fibres, which strengthen the plant. Root cells have thread-like outgrowths called root hairs, which absorb water and mineral salts.

Chloroplasts, the transparent cuticle and hairs show on a photomicrograph of a plant leaf

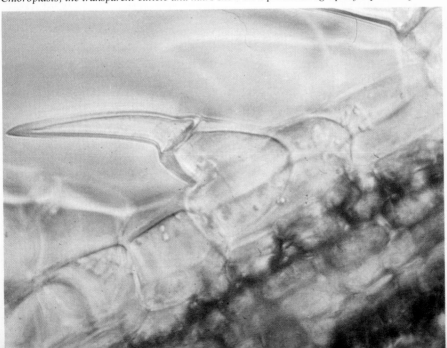

The most important part of a cell is its nucleus. We know this because the cell dies if its nucleus is removed from it. The nucleus is surrounded by a firm membrane, the nuclear membrane, and contains a material called DNA (deoxyribonucleic acid), the substance on which all life is based. DNA not only controls all the activities that occur in the cell but it is also the material that is passed on from one generation to the next and is responsible for likenesses between parents and their offspring. It is passed on in the form of chromosomes. This important role is possible because DNA is the only living substance that can make a copy of itself (replicate). A copy is passed from parents to their offspring and contains a 'pattern' which tells the new animal or plant how to develop. In sexual reproduction the offspring inherit one copy of DNA from the mother and another from the father, which is why offspring inherit characteristics from both parents.

Every DNA unit consists of two long chains each composed of alternate sugar and phosphate molecules. These chains are

An atomic model, to scale, of a DNA molecule

wound around each other in a spiral and are linked together by four important chemicals arranged in pairs, like the rungs linking two sides of a spiral staircase. This structure is called a double helix. The four chemicals can be arranged in many different ways in the double helix and can act in the same way as a code. The different chemical arrangements send different messages, which produce all the different characteristics of animals and plants.

When the DNA molecule replicates, the two chains separate from each other as if they were being unzipped and each chain then forms a new double helix by attracting the necessary chemicals to it. DNA also makes a smaller nucleic acid called RNA (ribonucleic acid) which acts as a messenger to carry the 'pattern' from the nucleus to the cytoplasm where it controls the formation of proteins, especially enzymes. When the newly formed RNA molecule passes into the cytoplasm it attaches itself to the ribosomes – the protein factories of the cell – on the endoplasmic reticulum.

The double helix structure of a DNA molecule

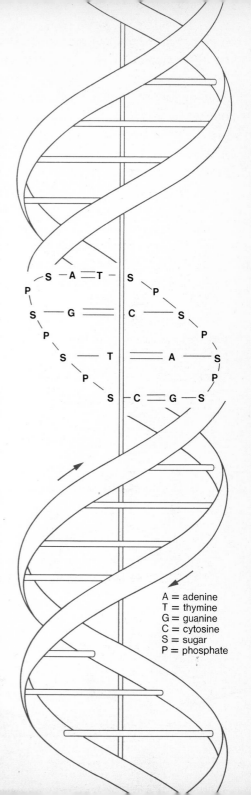

A = adenine
T = thymine
G = guanine
C = cytosine
S = sugar
P = phosphate

Cells: the smallest units of life
6   the nucleus and its chromosomes

The nucleus of a cell usually looks like a dense, roughly spherical body but when it is about to divide, thread-like structures (chromosomes) are visible. All cells except eggs and sperm contain two identical sets of chromosomes, one set from each parent of the organism. The number of chromosomes is fixed for each species: peas have seven pairs, humans 23, wheat 21 and crayfish 100. Egg and sperm cells have half the normal number of chromosomes – only one set.

The chromosomes consist of chains of DNA, with the genes, which control characteristics of the organism, arranged in order along them. All the cells of an organism are constantly being replaced and added to during the normal course of development and growth. Each cell divides into two new ones by a process called mitosis. During mitosis each chromosome divides into two so that two new sets are formed. One set goes to each end of the cell and forms a new nucleus. A membrane forms down the middle of the cell separating the two nuclei and so creates two new cells each with a complete set of chromosomes.

A fertilized egg (zygote) must have a complete set of chromosomes.

Human chromosomes magnified. Both males and females have 46 chromosomes, arranged in pairs, identical except for the sex chromosomes. The male sex chromosomes are the XY pair

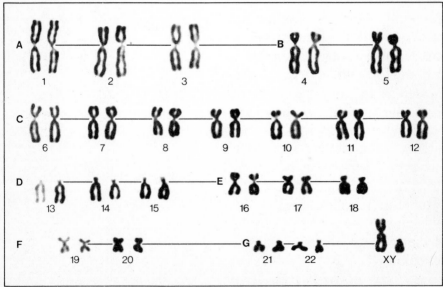

The egg and sperm of which the fertilized egg is made must each have half this number. In order to do this, eggs and sperm are produced by a special cell division called meiosis in which the chromosome number is halved. Thus the eggs and sperm of humans have 23 chromosomes, not 46. When the egg and sperm cells fuse they form a zygote having 46 chromosomes. During meiosis one chromosome set of the original cell goes into each newly-formed sex cell, but before they separate the two sets exchange hereditary material. This means that the offspring from these sex cells will differ in many small characteristics.

It is a small difference in one pair of chromosomes, the sex chromosomes, that determines whether an organism is male or female. In most female animals and plants the sex chromosomes are similar, and called XX, but in males they are different, and called XY. During meiosis one chromosome of each pair goes into the eggs and sperms. All the eggs will contain an X chromosome but half the sperm cells will have X and the other half Y chromosomes, so that there is an equal chance of male or female offspring being produced.

In the female the sex chromosomes are the XX pair. Note that the sets numbered 1 – 22 in both diagrams are the same – only the last sets differ

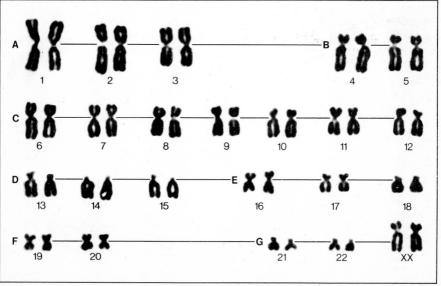

Evolution
7 Darwin and his theory

Until the middle of the 19th century it was generally believed that each species or type of animal and plant was separately created (Special Creation). The theory of evolution, on the other hand, proposed that all present-day species are descended by a process of gradual and continual change from animals and plants that have lived in the past.

Aristotle was the first to recognize that living organisms can be arranged in groups whose members show certain similarities. In the 18th century Linnaeus used this as the basis for his system of classification, dividing and subdividing animals and plants into groups all of which were thought to have arisen by Special Creation. However, the French biologists Lamarck and Buffon and the English naturalist Erasmus Darwin challenged this theory. They thought that evolution had taken place but could produce no evidence to prove it. Lamarck's theory of evolution proposed that characteristics acquired during an animal's lifetime could be passed on to its offspring. Charles Darwin, and another Englishman, Alfred Russel Wallace, working independently, produced the first convincing evidence about evolution. In 1858 they jointly proposed the theory of Natural Selection.

Compare the extinct sabre toothed tiger, trilobite and Cordaites, *with their modern relatives*

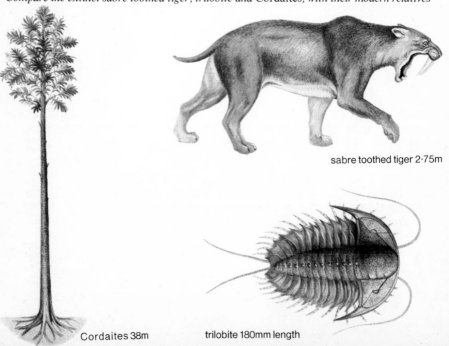

sabre toothed tiger 2·75m

Cordaites 38m   trilobite 180mm length

The following year Darwin published his famous book, *The Origin of Species*, in which he applied the theory of Natural Selection to the evolution of plant and animal species. This theory is based on the observation that the individuals of a species are not all identical but differ in many characteristics, such as size and eye colour. Many of these characteristics are hereditary, that is they are passed on from parents to their offspring. Most organisms produce large numbers of offspring but the size of the populations tends to remain the same. Therefore some individuals perish while others have characteristics that make them better able to survive and produce offspring themselves. These individuals and their offspring are said to be better adapted to their environment. Over a long period of time the conditions of the environment change, and the animals and plants also change as they adapt to the new conditions. In this way new species evolve from the old by a gradual process of continual change. Darwin and Wallace provided evidence for evolution from the anatomy, geographical distribution and the fossils of animals and plants. Today most people accept Darwin's theory, modified by theories of genetics.

All living plants and animals have evolved from ancient extinct forms

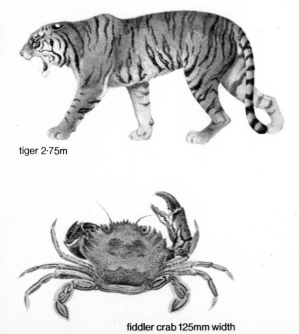

tiger 2·75m

fiddler crab 125mm width

Scots pine 30m

Evolution
8 evidence from living things

Some of the most important evidence to support Darwin's theory of evolution comes from the structure (anatomy) of living organisms and their embryos. The limbs of vertebrates, except fishes, are all built on the same basic plan, showing evidence of descent from a common ancestor. The hands of apes and man, the wings of bats and the flippers of whales all have the same basic bone structure. As Darwin said, 'the bones in their position and number are so similar that they can be classed and called by the same names'. The limb which is typical of these vertebrates is called a pentadactyl limb: basically it has five digits, as in man, although in many species this number is reduced. Limbs and organs of different animals which differ in function and superficial structure but which have the same basic structure are said to be homologous. All flowering plants, too, show evidence that they have evolved from a common ancestor. The tendrils of climbing plants and the spines of a cactus, although very dissimilar, have both evolved from the leaves of the ancestral plants.

Many animals have bones or organs which are of no apparent use to them, and are thought to be vestiges of organs that were useful to

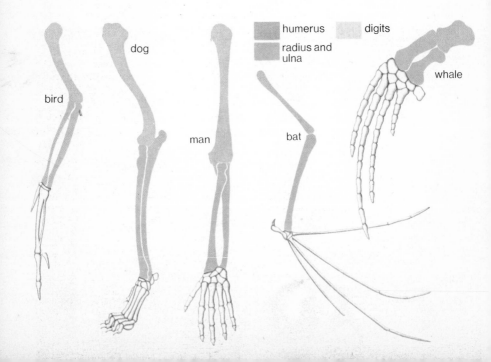

an ancestor at an earlier stage of evolution. All mammals evolved from land-living ancestors. The whale is an aquatic mammal whose forelimbs have become modified for swimming and whose hindlimbs have disappeared. When a whale is dissected, traces can be found of bones that are vestiges of the leg bones of its land-living ancestors.

In man there are many vestigial structures. We no longer have tails, but at the base of the human spine is an extra bone (coccyx) which is the vestige of the tail of our ancestors. The appendix is a large and useful organ in herbivorous animals such as sheep and cattle, and is probably involved in the digestion of cellulose. In man and other meat-eating mammals the appendix is still present but in a very much reduced form. This confirms that man and other mammals have evolved from a common ancestor.

The study of embryos also provides evidence to support the theory of evolution. The adults of vertebrates such as man, the chicken and the dogfish are very different from each other but their embryos are remarkably similar – for instance, they all have gill slits. This suggests that all vertebrates evolved from the same ancestor.

Left to right: *the embryos of man, chicken and dogfish are very similar*

Darwin collected much of the evidence to support his theory of evolution during his voyage around the world in the survey ship *Beagle* from 1831 to 1836.

Darwin visited two very similar groups of islands, the Cape Verde group off Africa and the Galapagos Islands off South America. Both are tropical and volcanic, but Darwin found that the Galapagos animals resembled those on the American mainland, while the Cape Verde animals were more like those of the African mainland. However, the fauna of both islands were slightly different from the mainland types. For instance, some enormous tortoises and an unusual type of marine iguana inhabited the Galapagos but were not found on the South American mainland. Darwin explained these differences by suggesting that the island animals evolved from the mainland forms which had managed to reach the islands and adapt to the particular conditions there.

Darwin was especially interested in the finches of the Galapagos Islands. They were clearly

Top: *marine iguanas of the Galapagos Islands.*
Below: *Galapagos tortoises weigh up to 450 kg*

related to the mainland finches, but had a greater variety of beak shapes and feeding habits. The mainland finches were mostly seed-eating birds with short stout bills. The island birds had evolved into types which fed on insects, seeds, cacti, grubs and fruit. Each island finch had a different type of bill, which was specialized according to the food it ate.

While he was travelling in South America Darwin saw animals that were clearly related to types found in the Old World but in certain ways were very different. For instance, the agoutis of South America are very like the rabbits of the Old World, but live in forests rather than open land. Rheas are flightless birds that resemble the African ostrich but are somewhat smaller. The South American jaguar is similar to the African leopard, and the llama is like a small camel. From these observations Darwin concluded that animals as they evolve adapt themselves to different environments. Their geographical distribution provides some of the most important evidence that species originated by evolution.

Drawings of the Galapagos finches show their specialized beaks and different colours

Geospiza strenua

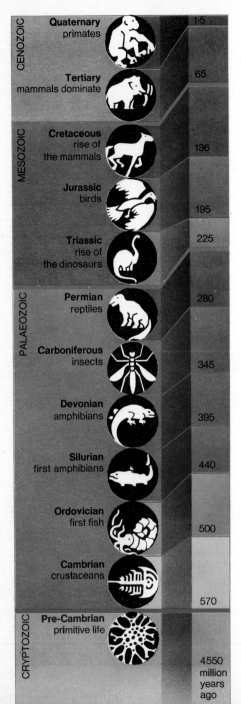

CENOZOIC	Quaternary primates		1·5
	Tertiary mammals dominate		65
MESOZOIC	Cretaceous rise of the mammals		136
	Jurassic birds		195
	Triassic rise of the dinosaurs		225
PALAEOZOIC	Permian reptiles		280
	Carboniferous insects		345
	Devonian amphibians		395
	Silurian first amphibians		440
	Ordovician first fish		500
	Cambrian crustaceans		570
CRYPTOZOIC	Pre-Cambrian primitive life		4550 million years ago

Fossils are traces or preserved remains of animals or plants that existed in the past, some of them millions of years ago. The most common fossils are shells, bones and teeth because it is unusual for a whole animal to be fossilized. The best-known example of an entire animal being preserved is the woolly mammoth, a type of extinct elephant whose whole body has been found in Siberia, frozen in ice from the time of the Ice Age of the Pleistocene. Most fossils are found embedded in rocks and are themselves impregnated with minerals so that they become hard and stony and last for many years. Until about 1800 fossils were curiosities, but during the 19th century scientists began to realize that fossils could provide valuable information about earlier animals and plants.

In 1830 Charles Lyell, a Scottish geologist, published a book called *Principles of Geology*. He explained that rocks are arranged in layers (strata), and that the lowest strata are the oldest. It is possible, therefore, to estimate the age of a fossil from the layer of rock in which it is found. Darwin took a copy of Lyell's book with him on his voyage in the

Beagle and it greatly influenced his theories of evolution.

In South America Darwin found fossils of a giant animal, *Glyptodon*, which was very similar to the modern armadillo. This discovery suggested to Darwin that the two forms were related and might have evolved from a common ancestor. Darwin also found fossil remains of an extinct giant sloth, *Megatherium*, which was similarly related to modern sloths. Since Darwin's time many more fossils have been discovered, and in general the fossils of less advanced animals have been found in the older rocks. For instance, primitive fishes have been found in Silurian rocks 400 000 000 years old, while remains of the primitive bird, *Archaeopteryx*, do not appear until much later, in Jurassic times. The fossil record is not complete but it does show evidence of a continuous and ever-changing succession of plants and animals. Many of the earlier types can be recognized as the ancestors of later forms which have evolved through the changing conditions of their environment.

Top: *fossil fish.* Centre: Archaeopteryx.
Below: *fossils found in a rock face in Iceland*

Evolution
11 evidence from horses and elephants

A complete series of fossils that show all stages in the evolution of a particular animal group is a very rare find. However, in North America there is a very detailed fossil record of the evolution of the horse over a period of 60 000 000 years from the Eocene period to recent times.

From these fossil bones zoologists traced the evolution of the horse from its earliest form, *Hyracotherium* (or *Eohippus*), to the present-day genus, *Equus* (horses, zebras, asses). As they evolved, horses gradually became larger, their teeth became more complex, and their toes were reduced in number. These changes came about with the change in the environment from forests to prairies. The earliest horse, *Eohippus*, was a small animal about the size of a fox terrier, and it lived in the forests. It had four toes on the front feet and three on the hind, giving a broad surface useful for walking over soft ground, and it had a short jaw and simple teeth suitable for feeding on forest trees and shrubs. In the Oligocene period it had evolved into a larger forest-living form, *Miohippus*, with fewer toes on the feet.

Evolution continued, with increases in size, until the modern horse,

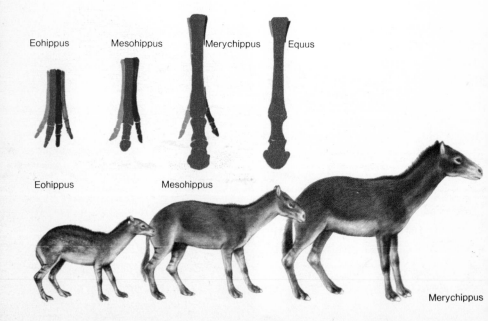

Eohippus Mesohippus Merychippus Equus

Eohippus Mesohippus

Merychippus

Equus, first appeared at the beginning of the Pleistocene period. *Equus* lives and grazes on open grasslands. The number of toes on each leg has been reduced to one, which forms the hoof, so that the modern horse is better adapted to galloping over dry open land to escape from its enemies.

The history of the elephant can be traced in a similar way, although there are far more fossil remains of the horse than of the elephant. The earliest elephant was about as big as a pig, and had very short tusks and a short snout. It lived in Egypt at about the same time as *Eohippus* lived in the forests of North America. Many types of elephant evolved from this animal. These types included the woolly mammoth which was very common during the Ice Age and whose remains have been found in northern Europe and Asia, North America and Africa. The woolly mammoth is one of the few animals whose remains have been completely preserved, because they were frozen in ice. Cave paintings made by Cro-Magnon Man more than 40 000 years ago show animals including the woolly mammoth. Today there are only the African and the Indian species.

ohippus

Equus

Evolution
12 from ape to man

All races of modern man belong to the same species, *Homo sapiens.*
They share unique characteristics such as a high degree of intelligence
associated with a large rounded skull, an upright posture and an
ability to communicate verbally. Until about a hundred years ago
very few people would accept the idea that man, like animals, had
originated by evolution from lower animals. Today it is generally
believed that modern man and modern apes both evolved from
common ancestors.

The earliest primates formed a separate evolutionary line from the
other primitive mammals about 70 000 000 years ago. About
50 000 000 years later ape-like animals of the genus *Proconsul* (*Dryo-
pithecus*) appeared in East Africa and it is thought that the higher
apes and man evolved from these primitive forms. The separation of
apes and man probably occurred between 3 000 000 and 15 000 000
years ago. At this time man-like primates of the genus *Australopithe-
cus* were living in southern and eastern Africa. From their fossil
remains they seem to have been near human in appearance and they
form a definite link between apes and man. They probably gave rise

British Museum (Natural History) reconstructions of Proconsul (left) *and Peking Man* (right)

to the earliest men: members of the genus *Homo*, including Peking Man and Java Man, both classified as *Homo erectus* because of their upright posture. They lived in Africa about a million years ago and also in Asia. They probably used fire and tools and lived in caves.

From these early types evolved other primitive men of the genus *Homo*. The earliest appeared in Europe about 300 000 years ago. One of the best known is Neanderthal Man (*Homo neanderthalensis*), a stockily-built type, about five feet tall, with a low forehead and rather short legs. He lived in caves in Europe between 150 000 and 35 000 years ago and his remains have been found in Germany, Gibraltar, North Africa and in northern Asia. He used fire and tools and buried his dead. Between 40 000 and 130 000 years ago the first representatives of modern man (*Homo sapiens*) appeared. One group, called Cro-Magnon Man, lived in France. They were tall, finely-built men who used flint tools, made clothes and did cave paintings, including those depicting the woolly mammoth, which can still be seen today. It is thought that modern races of man evolved from Cro-Magnon and similar groups in different parts of the world.

Left: *Neanderthal family group of the last Ice Age.* Right: *men of the Cro-Magnon type*

2   Tees-Water Old or Unimproved Breed

1   Tees-Water Improved Breed

Evolution
13   *how it happens*

Darwin provided a great deal of evidence in support of evolution and he proposed the theory of Natural Selection to explain how it could have taken place. In any population of animals or plants there are many variations, some of which are of more use than others in helping the individuals to survive. Because many of the characteristics of an organism are passed on from parents, future generations will also have the benefit of these characteristics.

In recent times, a striking example of evolution actually happening is seen in the development of the peppered moth (*Biston betularia*). The original colour of all populations of this moth was a pale, speckled grey. This colour blends in very well with the lichen on the tree trunks on which the moth lives, and effectively hides it from birds, its chief enemy. About 120 years ago a black variety suddenly occurred by chance. In industrial areas the tree trunks are covered with soot and so the black moths had a great advantage over the pale ones because they were less easily seen. Today, in industrial areas,

Top: *before and after selective breeding.*
Below: *original and mutant peppered moths*

the black form of the peppered moth is far more common than the pale form. In this way the black moth has evolved from a variation in the natural (pale) population and has increased because of its camouflaging colour.

Variations occur by chance in a natural population and are selected if they have any advantage over the normal forms. Another type of selection, however, has been practised by man for hundreds of years and provided Darwin with further evidence to support his theory. Certain characteristics in breeds of domestic animals, and crops, such as wheat, are more useful to man than the natural forms. He has, therefore, bred domestic animals and crops so that they have features that are not in natural populations. Improved breeds of sheep and cattle are examples of selective breeding.

Many varieties of ornamental and crop plants have been produced by breeding from wild species. For example, some cultivated roses have been derived from the wild dog rose.

Top: *wild roses.* Below: *'Waltztime', a cultivated rose bred from different hybrids*

△ parent plants

△ first generation
▽ second generation

Darwin observed that many variations existed in natural populations but could not explain how they occurred or how they were passed on from parents to offspring. In the 1860s Gregor Mendel, an Austrian monk, made some important discoveries about heredity.

Mendel studied the transmission of inherited characteristics in the garden pea plant. He selected plants with easily recognized characteristics: differences in size, colour of flowers and appearance of seeds. First he made sure that his peas always bred true for a particular characteristic, for example, that purple-flowered plants would produce only purple-flowered offspring. He then fertilized a purple-flowered pea plant with pollen from a white-flowered plant. All the plants from this cross had purple flowers. When these plants pollinated themselves he found that in the second generation there were almost exactly three times as many purple-flowered as white-flowered plants. He obtained similar results with other pairs of characteristics. In the second generation there were three times as many smooth as wrinkled peas, and three times as many tall as dwarf plants. From these results he came to several important conclusions.

Characteristics such as height, flower colour and texture of seed coat are determined by factors that must be present in the cells of all organisms and therefore are passed on to their offspring. It is now known that these factors, called genes, form parts of the chromosomes. The genes themselves pass unchanged from one generation to the next, which results in a particular characteristic of an organism being present in its offspring. Most organisms possess two similar sets of chromosomes and thus carry two factors for any one characteristic. Although the genes are passed unchanged from parent to offspring, Mendel also found that the characteristics of a particular gene could be altered in some way by another gene. Thus for each characteristic that Mendel studied, one of the factors, called the dominant factor, masked the effect of the other, the recessive factor. So the factor for purple flowers was dominant to that for white flowers, which was why all the first-generation plants had purple flowers. By the second generation the factors had separated and both purple and white flowers were produced.

The animals and plants we know today have evolved from earlier forms by a gradual and continuous process of change. Individuals in any population vary, and those having characteristics most suited to the conditions in which they live will survive, while those unable to adapt will die. Many characteristics are hereditary and are controlled by the genes which form part of the chromosomes of every cell. When two sex cells unite the chromosomes combine, and in this way characteristics are passed on from parents to their offspring. This is why individuals resemble one parent in some respects and the other parent in other ways.

How do variations in characteristics come about? Some are caused by a sudden change (mutation) in a gene. This happens when a mistake occurs during cell division and a gene is not copied exactly. The imperfect copy is called a mutant. Instead of producing the normal characteristic, the mutant gene will cause a variation. For example, the normal gene in the peppered moth produces a grey colour but its mutant produces a black coloration. On average, in most organisms, mutations occur in one out of every 500 000 genes. If a mutation

When a five-legged bull is born in Malaya, it is considered sacred

occurs in the sex cells it will affect the development of the offspring from those cells and will be inherited by future generations. These mutations occur because the DNA of the chromosomes is not stable and can be affected by certain external factors such as radioactivity. Mutations may also occur in the chromosomes themselves. Most normal plants and animals have two sets of chromosomes and are called diploids. Some plants, however, have more than two sets, a condition known as polyploidy. The chromosome sets may be multiplied by three (to produce a triploid organism), four (to give a tetraploid), and so on. The type of wheat used to make bread is thought to be a hexaploid (with six chromosome sets) which has been produced by the artificial mating of different species of wild diploid grasses.

Most mutations do not benefit the organism and can even produce harmful or grotesque results. In natural populations such mutations will die out. However, in some rare cases, these mutations have been deliberately preserved by man. In Malaya, for example, the unusual five-legged bulls are kept alive and bred because they are considered to be sacred.

Mutations in humans, such as six-fingered hands, are somewhat rare

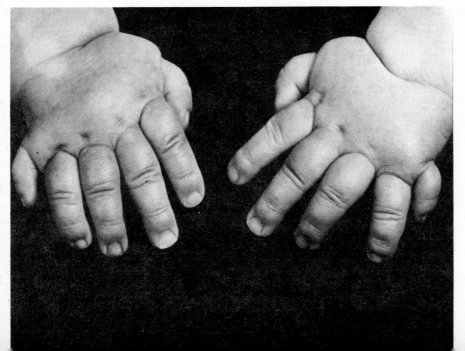

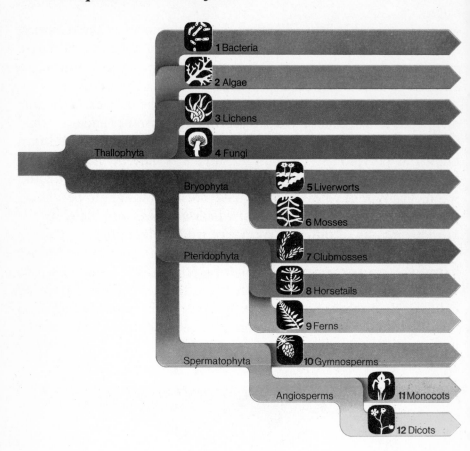

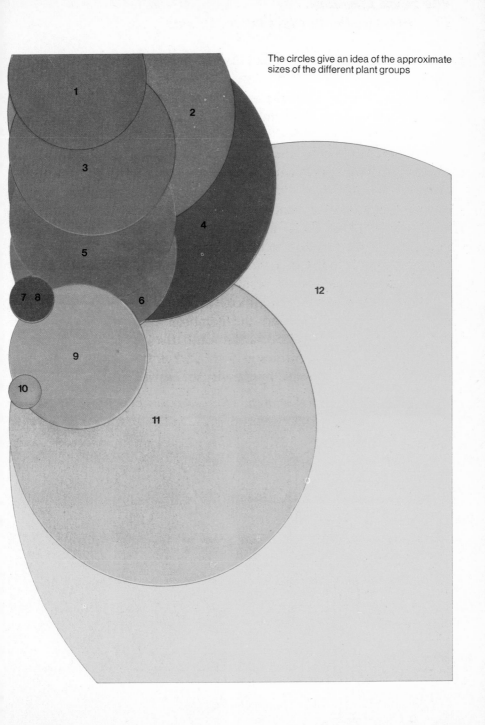

The circles give an idea of the approximate sizes of the different plant groups

Viruses are on the borderline between living and non-living things. Some have the properties of living organisms, but others do not. Viruses are smaller than the smallest bacteria and look like minute rods, spheres or spirals from 10 to 400 millionths of a millimetre in diameter. The largest are barely visible under the most powerful optical microscope so their structure has been mapped with the electron microscope. Because viruses are parasites they cannot grow or reproduce outside the living host cell that they invade. They consist of a protein coat which encloses a coil of hereditary material (DNA or RNA) and they have no nucleus or cytoplasm. The hereditary material is injected into the host and takes over the working of the host cell so that many new viruses are produced inside it. The host cell bursts to release the viruses which in turn can infect more cells.

In man viruses cause smallpox, measles, mumps, yellow fever, poliomyelitis, influenza and the common cold, all of which are infectious. Viruses cause foot-and-mouth disease and Rigs disease in cattle. They also cause rabies in dogs – when the dog goes mad and bites it spreads the disease by the viruses carried in its saliva. Many

Left: *construction of a bacteriophage.* Right: *bacteriophage photograph enlarged 450 000 times*

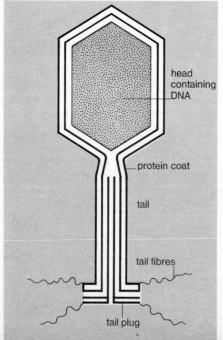
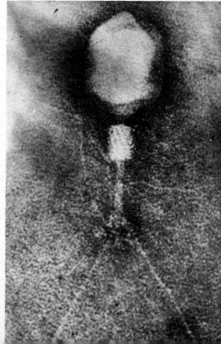

head
containing
DNA

protein coat

tail

tail fibres

tail plug

plants, such as potatoes, tobacco and tomatoes, suffer from virus diseases which deform or kill them, but the attractive yellow streaks on tulip petals are also caused by a virus. There are no drugs to cure virus diseases in the same way that antibiotics help to cure bacterial diseases. However, our bodies produce chemicals called antibodies which fight viruses. Also, doctors can produce substances called vaccines which consist of dead or less powerful viruses. These are injected and stimulate the body to produce antibodies more rapidly. In this way people can be inoculated against some of the virus diseases, including smallpox and poliomyelitis.

Man has made use of certain viruses to kill pests. For example, the virus disease myxomatosis was used to kill rabbits which caused great damage to farmers' crops. But the surviving rabbits eventually developed antibodies to fight the myxomatosis virus.

Viruses which attack bacteria are called bacteriophages (bacteria-eaters) or simply phages. They are the largest and most complicated viruses. Many have a tail for injecting DNA into the bacteria which disintegrate in about half an hour, releasing hundreds of new viruses.

Left: *tobacco mosaic virus magnified 180 000 times.* Right: *Hereford steer with Rigs disease*

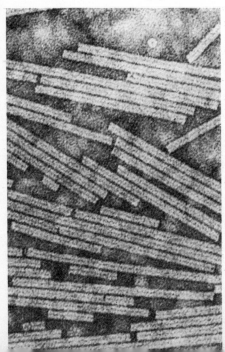 

Bacteria are very small single-celled organisms (micro-organisms) that exist in enormous numbers almost everywhere. They live in soil, water, air, and in living and dead animals and plants. A gram of soil can contain up to a thousand million bacteria, and there may be hundreds of thousands in a single drop of milk.

Bacteria differ from each other mainly in where and on what they live, and in the shape of their single cells. There are the spherical coccus types such as *Staphylococcus* and *Streptococcus*, which often occur in chains or masses, and the rod-shaped bacillus type such as *Mycobacterium*, which causes tuberculosis. Other disease-causing bacteria are *Eberthella typhi* (typhoid), and *Vibrio cholerae* (cholera). The type of bacterium which forms a coil or spiral is *Spirillum*.

Although bacteria cells are more complicated than viruses they are still very simple. Their structure has been worked out with optical microscopes which magnify by over a thousand times, and electron microscopes which magnify by one hundred thousand times. All bacteria have a tough outer cell wall so their food must be soluble before it can be absorbed into the cell. In some bacteria there is a

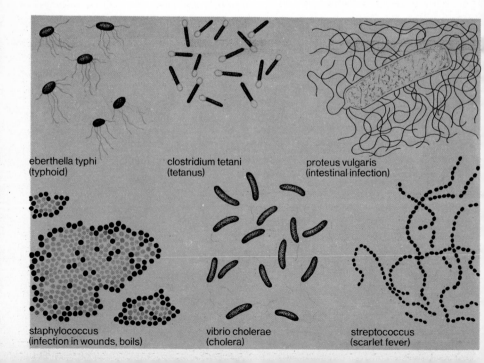

eberthella typhi
(typhoid)

clostridium tetani
(tetanus)

proteus vulgaris
(intestinal infection)

staphylococcus
(infection in wounds, boils)

vibrio cholerae
(cholera)

streptococcus
(scarlet fever)

protecting layer of jelly enclosing the cell wall, and also one or more minute fibres (flagella) used for swimming. Inside the cell there is a coil of DNA and other chemical substances, but there is no definite nucleus or any of the other structures found in plant and animal cells.

Bacteria usually reproduce by simply splitting in two. When temperature conditions are favourable, about 37°C for most bacteria, they can divide about once every 30 minutes. In theory, one bacterium could form about 140 000 000 000 000 bacteria at the end of 24 hours. In fact, this does not happen because the supply of food soon runs out, poisonous wastes accumulate, and after a time no more bacteria can survive. Although most bacteria reproduce by dividing in two, some bacteria can reproduce sexually, during which the contents of one bacterium flow into another.

Bacteria are very tough. Different kinds can live in almost every environment, from hot springs to arctic frost. Many can form a type of spore under certain conditions. A spore is a tiny capsule inside which a bacterium can survive for years in spite of drying out, intense heat or disinfectant. Few of the disease bacteria can make spores.

This jelly (agar) was inoculated with a type of bacterium which then grew and multiplied

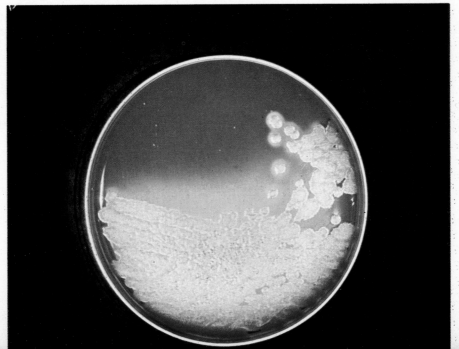

Many bacteria live as saprophytes, which means that they use dead material as their source of food. By doing this they carry out a function which is vital to all living things – they allow the supplies of oxygen, carbon dioxide and nitrogen compounds to be used again and again. Unless bacteria cause the rotting of dead material, plants obtain no food from the soil, and without vegetation animals cannot live.

Man made use of bacteria in making butter and cheese long before he knew that such organisms existed. They play an essential part in making cheese, yoghurt and vinegar, in the tanning of leather, in the preparation of linen from flax and in the treatment of sewage. Bacteria are also useful in keeping rivers and lakes clean.

Most bacteria, like higher plants and animals, need oxygen from the air to keep alive. A few bacteria use chlorophyll to build up their food as other green plants do, but most of them have alternative ways of finding their food and getting energy. Some use iron or nitrogen for their energy-producing processes, some (the saprophytes) live entirely on dead matter, and others are parasites, feeding on living plants and animals. Sulphur bacteria live in hot springs, mud and

Photomicrographs, taken at one-, three- and eight-hour intervals, show bacteria dividing

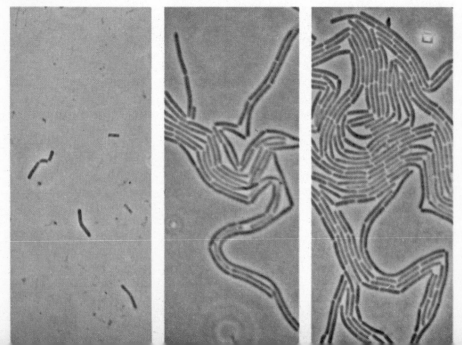

stagnant water and obtain their nourishment from the sulphur in the water. The bad-egg smell of these bacteria comes from hydrogen sulphide which they produce. Nitrogen bacteria, especially those that can turn nitrogen from the air into nitrates, are very important in farming. Some of these bacteria, such as *Rhizobium*, live in the roots of peas and clover.

There are some bacteria which live as parasites in plants and animals and cause diseases. They are called pathogenic (disease-causing) bacteria and we know them as germs. The bacteria that are parasites on man cause diseases such as diphtheria, typhoid, leprosy, tuberculosis, tetanus and plague. Unlike virus diseases, bacteria can be treated with antibiotic drugs as well as vaccines.

The poisonous wastes produced by the bacteria in rotting food can lead to stomach upsets. When milk is pasteurized it is heated to a high enough temperature to kill all the bacterial cells except the spores. As the disease-causing bacteria carried by milk cannot form spores, the milk is safe to drink. But because of the growth of the harmless bacteria from the spores the milk can go sour.

Leprosy, caused by a bacterium that is parasitic, results in malformations of bones and limbs

Fungi are much more complicated than viruses and bacteria. Almost all of them are composed of many cells, each of which has a nucleus, storage grains, cytoplasm, vacuoles and other distinct parts. Fungi have no roots, leaves or flowers and, unlike other plants, they have no chlorophyll and so cannot make food. Some fungi are saprophytes, obtaining their food from dead matter, and others are parasites, taking their food from living plants and animals.

Most fungi are made up of numerous very fine threads called hyphae which are used to push into their food. These hyphae may be massed together to form a large structure, as in a mushroom or toadstool. Fungi grow almost anywhere, especially in places where it is mild, damp and dark. They reproduce by means of minute structures called spores which are produced in their millions and dispersed by air, water or on the skins of animals. If conditions are unfavourable, spores can stay in a resting (dormant) stage for up to fifty years.

One of the common simple fungi is the black pin mould, *Mucor*, which appears on moist bread left uncovered for a day or two. Its whitish thread-like hyphae weave together to form a mycelium, a

Left: *asexual reproduction in the pin mould to produce spores.* Right: *sexual reproduction*

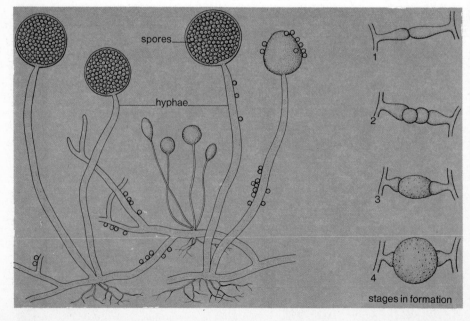

spores

hyphae

1

2

3

4

stages in formation

mass resembling cotton wool, which spreads throughout the bread. Some hyphae grow above the surface like tiny pins and are topped with black heads which are full of spores. When the spores are ripe the head bursts and scatters them. One of the most destructive simple fungi is *Phytophthora infestans* which causes widespread potato blight in crops which have not been protected by spraying. Dutch elm disease, apple scab and rose mildew are all caused by simple fungi.

There are many useful fungi. The green fungus that grows on mildewed oranges and lemons and on bread belongs to the genus *Penicillium*, from which the antibiotic drug penicillin was extracted. Yeast fungi are important to man. They are used in making wine, spirits, beer and in baking to make bread rise. Yeast is a single-celled fungus consisting of oval cells which grow by producing buds (smaller cells). It causes sugar to be converted into alcohol, but only when no air is present. This conversion is useful in the brewing industry.

Saprophytic fungi in the soil, in the same way as soil bacteria, break down dead plants and animals into the chemicals of which they are made, and these chemicals enrich the soil, making it more fertile.

Under a microscope. Left: Penicillium *creates spores.* Right: *a yeast species produces buds*

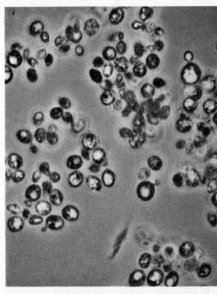

The plant kingdom
21 many kinds of fungi

The cultivated mushroom, *Psalliota campestris*, often used in cooking, is the part of the plant that grows above ground and produces the spores. In common with all fungi it consists of hyphae, which are woven into a stalk (stipe) and cap (pileus). Underneath the pileus are gills which radiate from the centre like spokes. The spores grow on these gills. They are pink at first, and then turn brown as the spores ripen. An average-sized mushroom can produce about a million spores every two minutes. It goes on producing these for two or three days. The underground part of the mushroom is the mycelium and consists of a looser network of hyphae which grow in soil, rotting dead plants or in manure.

There are several other edible fungi, such as chanterelles, boletuses and morels. There are also, however, many poisonous fungi, some having a distinctive colouring – the fly agaric, for example, has a white-spotted red cap and is very poisonous. Many other poisonous fungi look like edible ones, so wild fungi should never

Top: *bracket fungus*. Centre: *common field mushroom*. Below: *poisonous fly agaric*

be eaten unless they have been carefully identified. Other common fungi are bracket fungi, which grow from the sides of tree trunks and have holes instead of gills underneath.

Many kinds of complex fungi cause great damage to crops and buildings. Huge quantities of cereals have to be destroyed every year when they are attacked by such fungi as the rusts and smuts. The spores of these fungi blacken or redden the ears of cereals. The black rust of wheat, *Puccinia graminis*, common in the United States, is an example of this type of fungus. Many complex fungi are able to rot wood. One of the most destructive is dry rot, which can spread from wet wood into dry wood and infest a whole building. Several other fungi attack and kill living trees.

However, many of the toadstools and other fungi found in woods are useful. Some have underground hyphae that cause dead leaves and branches to rot. Others, called mycorrhiza, penetrate the roots of some trees and supply them with minerals from the soil.

When yellow rust fungus attacks cereal crops the leaves yellow and the plant dies

The plant kingdom
22   algae: the simplest green plants

The main difference between green plants and other living things is their ability to make food – complex starch and sugar molecules from simple carbon dioxide and water. They do this by the process of photosynthesis which requires the presence of chlorophyll, the material that produces the colour in all green plants.

Algae are the simplest organisms in the plant kingdom that possess chlorophyll and can make their own food. Algae vary in size: they may consist of a single cell, or a row of cells joined together to form a filament. The largest algae consist of a plant body called a thallus, which is made of many cells and gives the plant a ribbon-like appearance. Algae live in both fresh and salt water and may swim, float or fasten themselves to a firm surface. Some of them live in the soil and a few are parasites on animals and plants. All algae possess chlorophyll, but their green colour may be masked by other pigments. This provides a way of classifying them according to colour, for instance green algae (Chlorophyceae), brown algae (Phaeophyceae) and red algae (Rhodophyceae). Algae reproduce in various ways. Sexual

Photomicrographs. Left: *diatoms form an important part of plankton.*
Right: Spirogyra, *a thread-like green alga found in fresh water*

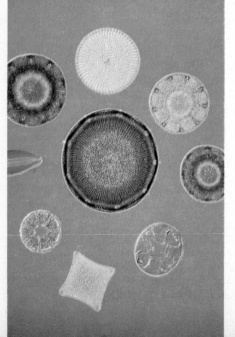 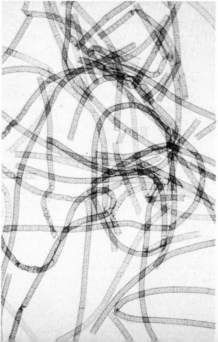

reproduction may occur, resulting in the formation of a new individual by the fusion of two sex cells. Sometimes asexual reproduction takes place, in which spores are produced, or a single-celled alga splits in two. Vegetative reproduction occurs when a filament or thallus breaks into two parts to form new plant bodies.

One of the most common of the single-celled green algae is *Pleurococcus*, whose colonies form a green film over the shady, moist side of tree trunks. *Spirogyra* is a common filamentous green alga which grows in ponds and stagnant pools. A salt-water alga with a thallus body is the sea lettuce, *Ulva*, which is eaten in salads in some parts of the world. Diatoms are single-celled algae (Bacillariophyceae), that live in fresh and salt water. They grow in enormous numbers and water-supply works often have to filter off tons of them every day. In salt water they make up a great part of the floating community of organisms called plankton, which is found in the upper layers of all seas and is the food supply for many aquatic animals. A diatom cell is built like a shallow box with a neatly-fitting lid and is impregnated with silica, which forms beautiful patterns.

Many kinds of green algae can be found in the stagnant water of ponds and pools

The best-known algae are the seaweeds – Phaeophyceae or brown algae – which are a common sight, especially along the shores of the Atlantic Ocean. The bladderwrack, *Fucus vesiculosus*, grows by the sea-shore on rocks exposed at low tide. It has a bushy, slippery, greenish-brown thallus that contains air bladders which act as floats, and it spreads out from a thick main stalk, growing up from a root-like base which anchors it to the rock. Some seaweeds are useful. Kelp is a large coarse seaweed which is burned to ashes and used as a source of iodine. The related oarweed, *Laminaria digitata*, is used as cattle fodder in Scandinavia and also yields a substance for thickening food. The young stems are cooked as a vegetable in Scotland and Ireland. *Sargassum* is a free-floating, brown seaweed that grows in dense masses in the Sargasso Sea, and *Macrocystis*, a giant plant, up to fifty metres long, forms thick forests in the Pacific Ocean.

The blue-green algae, Cyanophyceae, are primitive types, in many ways more like bacteria than algae. They usually form very small slimy clumps or strands, and are mainly found in hot springs where few other plants could survive. The first plants that carried out photo-synthesis probably resembled them. Some blue-green algae form associations with fungi, producing lichens.

Red algae, Rhodophyceae, vary in size from a few centimetres to one or two metres in length. Some of the Rhodophyceae, such as *Corallina*, do not look red at all because they are encrusted with lime. *Corallina* lives in close association with corals and helps to build coral reefs and atolls. *Delessaria sanguinea* is one of the showiest red algae. Its thallus consists of a stem with purplish-red fronds that look very much like leaves. The red pigment helps to absorb the limited amount of light in deeper water which enables the red algae to live at lower sea levels than the green algae. Some red algae are useful to man. In Japan, certain kinds are cultivated for a food called nori, and in Northern Ireland *Chondrus crispus* is made into a pudding. Agar, the jelly-like substance on which biologists grow bacteria and fungi cultures, is prepared from red algae.

Bladderwrack, a common seaweed, grows on rocks in salt water. Its bladders, filled with air, explode when pricked. Reproduction is by means of spores at the tips of the branches

Lichens are primitive plants that grow in thin flaky crusts, usually on walls or bare rock where very few other plants could survive. A lichen is not a single plant at all, but a combination of one particular species of fungus living together with one particular species of alga. Looked at under a microscope, the lichen plant body (thallus) is made up of a tangle of fungus threads (a mycelium) with a single-celled green or blue-green alga scattered through it. No fungus can feed on a bare dry rock and no alga can survive without moisture. Living together, the fungus holds moisture in its mycelium and the alga makes food which supports the fungus. This relationship between two types of organisms living together, to the benefit of both, is called symbiosis.

Symbiosis allows lichens to survive in very harsh climates. In the antarctic there are more lichens than any other kinds of plants, and in the far north a lichen called reindeer moss, *Cladonia rangiferina*, is the pasture food of the Laplanders' reindeer herds. Most lichens grow in bright sunlight, which the algae need to make food. The common wall lichen, *Xanthoria parietina*, attaches itself to walls

Reindeer moss, a type of lichen, is an important food for reindeer. It is one of the few plants that can grow in extreme northern climates

by its fungal hyphae; it forms a grey crust in wet weather and is yellowish in dry weather.

Lichens grow very slowly and some take as long as ten years to increase their size by a single millimetre. By measuring the thallus crust from the centre it is possible to calculate the age of a lichen – some are estimated to be over four thousand years old.

As well as the crust-like lichens, there are some that grow little stems like goblets or trumpets; *Cladonia pyxidata* is one of these. The biggest lichen plant is longbeard, *Usnea longissima*, which hangs like strands of hair from the branches of trees. Lichens were once used to make dyes, especially in Scotland for dyeing Harris tweeds.

Litmus, the chemical indicator that turns red with acids and blue with alkalis, is made from lichens growing in Holland and South America. Lichens can also indicate the amount of pollution in the air. When pollution increases to a certain level it kills many kinds of lichens, which is why they are not usually found in areas where there is a great concentration of industrial waste in the air.

Left: *an attractive lichen that grows on brick or stone walls.* Right: *longbeard lichen hangs from tree branches in the tropics. It is very different from its crust-like relatives*

Mosses and liverworts make up the group Bryophyta, all of which reproduce by spores rather than seeds. Bryophytes are small plants at an evolutionary stage intermediate between life in water and life on dry land. They must be surrounded by water, which prevents them from drying out and is also used for their reproductive process. They are therefore all found in moist habitats.

Liverworts usually grow in the form of a flat plate of cells against the ground, anchored by hair-like root structures (rhizoids). *Pellia* is a common kind of liverwort with a typical method of reproduction. Male and female sex organs are produced on the same plant body (thallus). The male sex organs are formed on the upper surface of the thallus and produce sperms (male gametes). The female sex organs are formed at the tips of the plant and produce the eggs (female gametes). Because the thallus produces the gametes it is called the gametophyte. A sperm swims to an egg in the water around the plant, and fertilizes it, forming a zygote, which remains on the gametophyte and grows to form a stalk and a capsule. The capsule produces many

Ground-clinging liverwort and moss, both bryophytes, in the same stages of reproduction. Spores are produced in the capsules at the tips of the stalks

spores and therefore the capsule and stalk together are called the sporophyte. The spores are spread by wind and grow into new gametophytes which, when mature, produce sex organs. The gametophyte and the sporophyte are produced alternately; this is called alternation of generations and it also occurs in mosses and pteridophytes.

A typical moss is *Funaria* which has leaves and stems that resemble higher plants, but no veins and no cuticle to prevent the plant from drying out. In some mosses such as *Polytrichum* (hairy cap moss), there is a central core of longish cells which gives extra strength to the stem. Unlike liverworts, many mosses grow as separate male and female plants. A tuft of moss consists of many tiny individual plants. Each plant consists of a rosette of leaves, at the centre of which the reproductive structure can be seen as a reddish dot. The process of reproduction is very similar to that of liverworts. Most mosses grow on trees, walls or between paving stones. Sphagnum is found on hillsides and moorlands, and its partially decayed remains solidify to form peat.

Sphagnum moss with its spore capsules. Dead sphagnum moss forms peat, which is used as a fuel and also to improve the texture of garden soils

The plants classified as pteridophytes are the horsetails, clubmosses (lycopods) and ferns.
They all reproduce by spores (tiny
one-celled structures) rather than
by seeds. They have true stems,
leaves and roots, linked together
inside by a conducting (vascular)
system, which carries water and
sugars within the plant and is also
the main means of support. All
land plants except the smaller
bryophytes possess vascular systems, part of which can be seen as
the veins in the leaves. The vascular system consists of phloem
cells, which carry solutions of
food made by the leaves to the
stem, roots and growing tips, and
xylem cells, which conduct water
from the roots to the rest of the
plant. Land plants also have a cuticle covering their leaves which
prevents them from drying out.
Clubmosses are found throughout the world, especially in wet
tropical areas. Some grow on
moist mountain slopes in cooler
regions, and the resurrection
plant, *Selaginella*, can survive
long droughts. In arid conditions
it forms dry brown masses resembling tumbleweed, but when in
contact with water, it quickly

An enlarged spore-producing cone of a horsetail

opens up and becomes green. Clubmosses have spirally-arranged leaves and most have a cone-like structure that produces the spores for reproduction. The fossil clubmosses included *Lepidodendron*, nearly 40 metres high, and *Sigillaria*, with a trunk more than two metres wide.

The horsetails, such as *Equisetum*, are also survivors of a formerly common and widespread group, and are still found everywhere except in Australia. They are most common in poor, sandy or marshy soils. Horsetails have an underground stem (rhizome) which produces up-right jointed stems with small wedge-shaped leaves arranged in circles at intervals around the stem. At the tips of the branches are spore-bearing cones and the plant is stiffened by grains of silica (sand). Few modern horsetails grow to a height of more than a metre, but fossils such as *Calamites*, which were once abundant in coal forests, grew up to 30 metres high. It is thought that most coal was formed from remains of these tree-like horsetails and clubmosses, which lived in estuaries and deltas. The dead plants became partly decayed and compressed in a process similar to the formation of peat.

Left: *clubmosses grow throughout the world.* Right: *a common horsetail (16th-century engraving)*

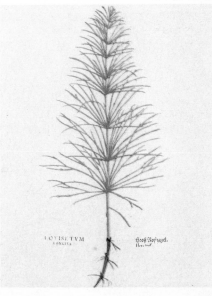

Ferns are pteridophytes, and form the largest group of vascular plants without flowers or seeds. Many have big feathery leaves (fronds) which unroll from the tip as they grow. Spores for reproduction are produced on the undersurface of certain of the leaves (pinnae). Bracken (*Pteridium*) is a common example which grows widely in dry places. Most ferns grow in moister places than bracken; the male fern, *Dryopteris*, is commonly planted in gardens or found in woodlands.

Many ferns vary from the typical ones. Hart's tongue, for instance, has long undivided leaves, and some small ferns, such as the spleen-worts, grow in the damp crevices of walls and cliffs. Giant tree-ferns, such as *Dicksonia* and *Cyathea*, grow in Australia, New Zealand and southern Pacific islands, often reaching a height of 20 metres. Some ferns live in or on water. *Marsilea* produces large hairy leaves above the water surface. *Azolla* actually floats on the surface, looking like a cross between duckweed and a liverwort.

In late summer a fern produces rows or groups of brown dots (sori) on the undersurface of some of its leaves on either side of the mid-

Spores on the underside of male fern leaflets. Only some of the leaves can produce spores – the others are used only for photosynthesis

veins. Each sorus consists of a group of sporangia (spore cases), with each sporangium containing several hundred spores. When the sporangium is ripe the cells which form its outer layer dry out, causing the whole sporangium to split open and throw the spores into the air. The spores, light as dust, are carried away in air currents.

The spores must land on a moist medium, otherwise they will not grow. Each spore develops into a fingernail-sized plate of green tissue (prothallus) bearing male and female sex organs, and is attached to the soil by very fine roots (rhizoids). The prothallus does not grow into a fern; it is short-lived and its use is simply to produce sperms and eggs. When the sperms are released, they swim in the surrounding moisture until they come in contact with an egg. The fertilized egg (zygote) grows into the new fern plant. The prothallus is called the gametophyte because it bears the eggs and sperms (gametes). The fern itself is called the sporophyte because it bears the spores. Many other plants, especially the bryophytes, produce gametophytes and sporophytes: the process is called alternation of generations.

Left: *a drawing of the male fern shows the contrast between the under and upper surface of the leaves.* Right: *a tropical fern, which often grows as high as a tree*

Gymnosperms have a food- and water-conducting system and produce seeds. Members of this plant division have grown on Earth for about 300 000 000 years; the other division of seed-bearing plants, the angiosperms, is only half as old. The main difference between angiosperms and gymnosperms is that only angiosperms have flowers. When angiosperm seeds develop in the flower heads they are enclosed in an ovary which develops into a fruit. In contrast, gymnosperm seeds are usually exposed on the scales of a cone, such as the pine cone.

Seeds are a more efficient means of reproduction than spores because they are more likely to survive and grow into a new plant. Spores are usually delicate single cells and only one spore out of millions may ever germinate. Seeds are tough, and the tiny plant embryo inside a seed has a store of food and a protective seed coat (testa).

Gymnosperms include the well-known conifers and three other less familiar groups: the cycads, ginkgos and Gnetales.

Top: *Mediterranean or stone pine.*
Centre: *Douglas fir.* Below: *cedar of Lebanon*

Cycads are primitive fern-like plants, native to the tropics. They look like palms, with thick un-branched trunks crowned with fronds, and grow to a height of two to eight metres. The food sago is obtained from the stems of some species.

The ginkgo tree, *Ginkgo biloba*, is the last survivor of an ancient group. It is probably the oldest living plant species; fossils of an apparently identical tree go back to the end of the Carboniferous age, about 270 000 000 years ago. There are separate male and female trees. The seed-bearing females are not popular because their decaying seed coats are slippery and smelly.

Gnetales look a little like flowering plants. The most inter-esting is *Welwitschia*, of the south-western African desert. Its woody, carrot-shaped root grows down to great depths, but its stem is only a few centimetres high and swells out to a flat top over a metre across. The top consists of two long leathery leaves which tear lengthwise into ribbons and its cones look like primitive flowers.

Top: *ginkgo leaves*. Centre: Welwitschia *(fe-male)*. Below: *cycads at Kew Gardens*

The best-known gymnosperms are the conifers; the modern species are grouped into five subdivisions. Conifers are found throughout the world but some, such as the pines, usually grow only in the northern hemisphere, while others, such as the monkey puzzle tree, grow mainly south of the equator. Nearly all conifers are trees: the biggest of all is *Sequoiadendron giganteum*, the giant redwood or big tree of California, which can grow over 90 metres tall and reach an age of more than 4000 years. Most conifers are evergreens; they shed their leaves gradually so that the branches are never bare. The fallen leaves eventually decay and make the soil very acid, which is why few other plants can grow beneath coniferous trees.

Pines have clusters of needle-like leaves, and male and female cones on the same tree. Male cones are small, upright and yellowish, and occur in groups. In spring and early summer they release clouds of yellow pollen. Female cones are much larger, borne singly, and are made up of scales. Each scale has two egg-producing organs (ovules) on its upper surface. After pollen has blown onto the open scales of a female cone, the whole cone closes. Several months pass before the

Junipers are shorter than other conifers. **Right:** *yew seeds and leaves are poisonous*

sperm cell actually fuses with the egg cell. After fertilization the seed takes over a year to ripen. During this time the cone grows big and woody and eventually its scales open once again and the winged seeds blow away. Larches are the only common conifers to lose their needles in winter although their small round cones often remain on the tree for several years. Yews and junipers produce female cones which when ripe resemble berries. The female yew cone is a fleshy red cup called an aril with a single seed. Juniper cones are dark blue and fleshy.

Conifer wood makes valuable timber called softwood to distinguish it from the harder wood of angiosperm trees. European redwood or red deal comes from the Scots pine, *Pinus sylvestris*. White wood comes from the Christmas tree or Norway spruce, *Picea abies*, and another useful wood is derived from the western red cedar, *Thuja plicata*. The wood is easy to work and resists attack by insects, fungi and damp because it contains a protective resin. Much of it is converted to wood pulp for paper-making. Most forests planted in Britain today are of conifers such as Scots pine or the North American Douglas fir which grow more quickly than trees such as oak.

Scots pine is an important timber tree. Right: *redwood, one of the world's tallest trees*

runner bean

seed

nodules

There are more species of angio-sperms (flowering plants) than of any other division in the whole plant kingdom. They are relative newcomers, making their first appearance about 160 000 000 years ago. Before that time no flowers were known to exist. How and why they developed is still a great mystery because fossil evidence is still incomplete. How-ever, pollen grains found in beds of coal are a clue to when flower-ing plants began to grow. Once started, the group flourished.

Angiosperms are vascular plants whose seeds develop with-in an ovary. When this is ripe it is called a fruit, which can be fleshy, as in apples and pears, woody as in nuts, or dry and papery as in sycamore keys. Within the seed are one or two cotyledons (seed leaves) and the embryo plant. In some plants, food for the embryo is stored in the cotyledons, which then become swollen as in the pea and broad bean.

Botanists make a general clas-sification of the angiosperms into two groups: the monocotyledons (monocots) having one cotyledon and the dicotyledons (dicots) having two.

If a dicot such as a pea or bean

is taken from its pod and the seed coat removed, it falls into two segments which are the cotyledons. The monocots, on the other hand, have only one cotyledon but their structure is more complicated. A ripe grain of wheat or Indian corn (maize) is a fruit containing only one seed. Most of the seed consists of stored food, mainly starch, and the embryo is very small.

Seeds formed in summer usually germinate in the warm damp days of the next spring, but if they are kept dry they will still germinate after several years, although the period varies with the type of seed. In some plants the cotyledons form the first green leaves after the seed germinates.

There are other characteristics in addition to their cotyledons that distinguish the two groups. Dicot leaves are veined like a net, while monocot leaves are generally long and narrow with straight parallel veins. Dicot stems are often woody and some grow into tree trunks. Monocot stems are rarely woody: if they are, as in palms, the trunks are fibrous and not made of solid wood. The flowers of monocots and dicots are also different.

tiller

wheat

starch

seed

plumule

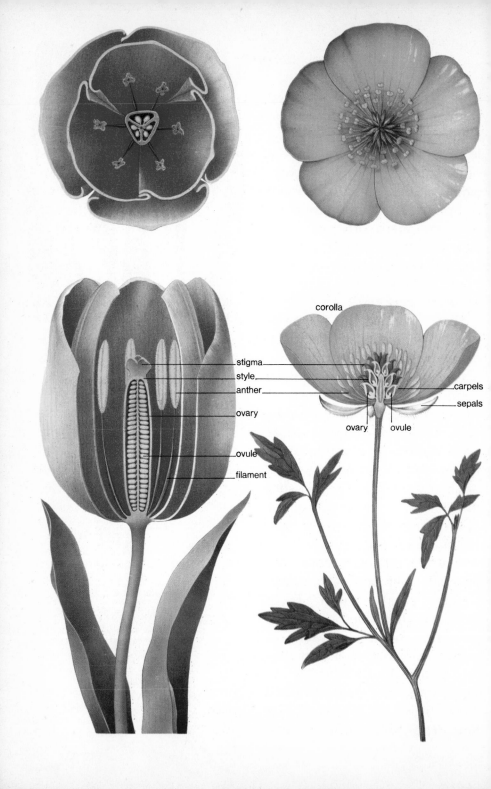

corolla

stigma
style
anther
ovary

ovule

filament

carpels
sepals

ovary   ovule

Botanists have several different ways of arranging angiosperms but they agree on the main division into dicots and monocots. These are further split up into plant families.

Typical monocots are grasses, cereals, lilies, tulips and orchids. Most other common flowering plants are dicots, including butter-cups, daisies, roses, cacti, most fruits and vegetables and most trees.

The whole purpose of a flower is to produce seeds, and its size, colour and scent all help to do the job. The smallest flowers belong to a pinhead-size plant, *Wolffia*, a type of duckweed that floats on the surface of small ponds. The biggest, almost a metre across, is the brown and purple *Rafflesia*, a tropical flower that looks like rotting meat and has an equally unpleasant smell.

Flowers are borne on special flower stalks, or pedicels, and the tip of the pedicel is swollen, forming the receptacle. Flower structure can be best understood by comparing two very simple and common plants – the tulip, a monocot, and the buttercup, a dicot.

The tulip has six coloured petal-like structures called perianth leaves; it does not have separate petals and sepals. The buttercup has an outer ring of five green sepals (calyx), which protect the bud, and an inner ring of five yellow petals (corolla). The brightly-coloured petals and perianth leaves attract insects and this brings about pollination. Inside the petals are the male structures, the stamens, each made up of an anther on a filament (stalk). Each anther consists of two pollen sacs containing pollen. When the anthers are ripe, they split open to release the pollen.

The tulip has six stamens, the buttercup a large number. In the centre of the flower is the female part which is made up of carpels. In the buttercup these are many separate scaly green structures each con-taining one ovule. The tulip has three carpels fused together into a swollen green structure, the ovary. The carpels and ovary both bear a short, stalk-like style, topped by a sticky-lobed stigma on which pollen lands. Inside the ovary there are three compartments represent-ing the carpels, and in each there are many small colourless ovules. These develop into seeds after fertilization, and the ovary of the tulip and carpels of the buttercup develop into dry brown fruits.

The most primitive of all the dicots belong to the family Magnoliaceae. The plants in this family are woody shrubs and trees, some of which are deciduous (shedding their leaves in winter) and some are evergreens. The characteristic big goblet-shaped flower has its parts spirally arranged. Most magnolias grow in north temperate and tropical regions and the prize cultivated ones have come from Chinese and Japanese stock.

The family Rosaceae consists of trees, shrubs and herbaceous (non-woody) plants and, although members of the family are widely distributed, most of them grow in north temperate regions. The family includes a great number of the fruits we eat as well as the genus *Rosa* to which cultivated roses belong. Cherry, plum, peach and apricot trees all belong to the genus *Prunus*, raspberry, blackberry, loganberry belong to the *Rubus* genus, apple is *Malus* and pear is *Pyrus*.

Native wild roses are found throughout the northern hemisphere. Common types are the wild pink dog rose and sweet briar; there are also tall luxuriant species which are found in the Indian and Chinese

Magnolias are popular ornamental trees for parks and gardens. Right: *the early-flowering wild cherry tree, whose fruit is not usually eaten, is a forecast of spring*

jungles. Cultivated hybrid tea roses have been produced in the last hundred years by artificial selection and breeding.

Another widespread family is the Leguminosae (legumes), of which peas, beans, clover, gorse and lupin are common examples. They all have butterfly- or keel-shaped flowers and pod-shaped fruits. Legume vegetables have been grown for food by man since as long ago as the Iron age. In more recent times they have also become useful in improving soil fertility. This is because legumes, unlike most plants, do not rely on nitrates dissolved in the soil. The bacteria living in their roots convert nitrogen in the air into a form which the plants can use.

We eat the unripe pods and seeds of runner beans, the unripe seeds in broad beans and peas, and the dried seeds of lentils and haricot beans. The familiar baked beans are prepared from haricot beans.

Perhaps the two most important legumes are the groundnut (peanut), from which edible oil is obtained, and the soya bean, which is very rich in oil and protein. Ground soya beans are used in a wide range of foods – from sausages to biscuits.

Spanish broom often grows in the wild. Right: *pea flowers in which the petals have withered and are about to fall off. The green pod becomes enlarged after fertilization takes place*

The cactus family, Cactaceae, comes from the New World, but cacti are now planted in many other countries and have even become pests. The prickly pear, *Opuntia*, made large areas of Queensland uninhabitable for a time. Cacti are mainly desert plants but have spread into tropical forests where, for instance, the Christmas cactus, *Zygocactus*, lives in the branches of trees. Cacti differ considerably in size and shape, but in typical examples their leaves have been reduced to spines and their stems are fleshy and green, or sometimes woody. One of the tallest is *Carnegiea gigantea* from Arizona, which is in the form of cylindrical columns up to 15 metres high. Two small spiny types which flower well are *Rebutia* from South America and *Mammillaria* from Mexico. The largest, most scented flower is produced by the forest-dwelling cactus, *Selenicereus*, whose flowers open at night. Epiphyllums (orchid cacti) have large scented orange flowers.

Heather, ling (heath) and rhododendrons all belong to the family Ericaceae, many of which

Top: *prickly pear cactus is a desert plant.*
Below: *water crowfoot grows in marshy areas*

live in moorland or mountainous areas. Most of the Ericaceae family are shrubs but a few are trees and there are also some smaller herbaceous plants. The many brilliant red, purple, yellow and orange azaleas have all been bred from the Japanese rhododendron since late in the 17th century.

The buttercup family, Ranunculaceae, is mainly herbaceous and common everywhere but in the tropics. Several members, such as the marsh marigold and water crowfoot, flourish in damp places. The group includes many beautiful garden plants, such as anemones, delphiniums, clematis and columbine.

Most of the family Cruciferae are herbaceous plants. They take their name from the cruciform (cross-shaped) arrangement of their four petals. Many have a sharp taste, such as mustard, radish, watercress and garden cress. The most important genus to man is *Brassica*, which includes the cabbage, cauliflower, broccoli, turnip and brussel sprout. Other attractive crucifers are wallflowers and aubrietia.

Top: *bell heather flourishes in Scotland.*
Below: *wallflower is a popular garden plant*

The family Umbelliferae are mostly herbaceous plants, with flowers arranged on a head somewhat like an umbrella turned inside out. Their stems are often grooved and contain a great deal of soft pith. They flourish mainly in temperate climates and some, such as cow parsley and hogweed, grow wild along country roads. There are also types that grow on mountains in tropical regions. Many plants of the Umbelliferae family have a distinct scent due to the presence of aromatic oils, and many are cultivated for food. Well-known examples are the carrot and parsnip, whose roots become swollen with starch and are eaten as vegetables, and parsley, whose leaves are used in cooking to add flavour. Others are dangerously poisonous, such as hemlock which is recognizable by its purple-spotted stems.

The family Labiatae includes a large number of plants with strongly-scented leaves such as mint, thyme, sage and rosemary. Many labiate leaves are used as herbs in cooking to flavour meat and vegetables. Lavender is a labiate whose flowers are dried and placed among clothes to keep them smelling sweet. Plants of the Labiatae family often have stems that are square-shaped when cut across, and

Two types of florets that make up the head of a daisy are ranged around the tip of the stem

flowers that tend to be small and two-lipped. They are found all over the world, particularly in the Mediterranean region.

The most advanced of the dicots is the Compositae family (composites) of herbaceous plants and shrubs. The flower head is not a single flower but a head (capitulum) composed of a number of small florets (little flowers). Each of the florets may have both male and female reproductive structures or those of only one sex. In plants such as the dandelion, all the florets on a single head are alike. However, in daisies those florets around the edge of the head (ray florets) each have a single, long, strap-like petal, while those at the centre (disc florets) have a very short, tubular corolla (fused petals).

The Compositae is one of the largest of all the angiosperm families. Lettuce, endive, chicory and artichoke are food composites; chrysanthemum, dahlia, zinnia and sunflower are decorative ones. Farmers and gardeners call the unwanted plants which occur among crops and flowers, weeds. Many weeds are members of the Compositae family, such as groundsel, thistle and dandelion, and they are distributed widely by means of parachutes of silky hairs attached to their seeds.

Hogweed is common in fields and hedgerows. Right: *mint is a popular flavouring for food*

Many timber trees are gymnosperms and produce softwood. All hardwood trees are dicots and many important ones, such as ebony, mahogany and teak, grow in the tropics.

Two of the most valuable hardwood trees of the temperate zone are oak and beech. The oak is one of the most important forest trees in temperate regions. All oaks produce fruits called acorns and most have leaves with wavy edges. Oak wood resists decay and was once the most important material for building houses and ships. One Mediterranean species, the cork oak, has an extremely thick bark which can be stripped off to provide cork. Beech is a very hard wood, often used to make furniture. Beech trees produce nuts as fruits and commonly form woods on chalk and limestone soils.

Both oak and beech trees live for hundreds of years and are pollinated by wind in the same way as many other trees. The pollen from the long catkins of the small male flowers is blown onto the female flowers which, after fertilization, develop into the fruits. Despite its name, the sweet chestnut, with its spiral-grooved trunk, is more closely related to the oak than to the horse chestnut.

Beech tree leaves change colour with the seasons. Right: cork oak tree bark is commercially useful

The sycamores and maples have leaves which look like a webbed hand. The sugar maple of the eastern United States and Canada is tapped for its sap to make maple sugar and maple syrup. The whole maple family is famous for its brilliantly-coloured autumn foliage. The maple is insect-pollinated and although the flowers are not large or showy they produce large quantities of nectar which attract bees. Ash and elm produce wind-pollinated flowers in early spring, and their fruits, like those of the sycamore, have large wings for dispersal by the wind. Elm wood does not rot easily and at one time was often used for underground water pipes.

Willow twigs are used in basketwork, and cricket bats are carved from the wood of one species. The small flowers, produced in tufts, attract insects to pollinate them in early spring when there is little other nectar available. Birch, hazel, poplar and aspen all rely on the wind to pollinate their catkins. The white-barked birch was used by the American Indians to make their canoes. Both birch and willow grow in climates much further north than other dicot trees – often in even harsher conditions than the hardy pine.

Left: *silver birch trees*. Right: *ash tree flowers, like all wind-pollinated species, are small*

Some of the most beautiful garden flowers, such as lilies, tulips and bluebells, belong to the lily family, Liliaceae. They flourish world-wide, and nearly all of them grow from bulbs, corms or rhizomes. Most tulip varieties (genus *Tulipa*) originated in central Asia, and were brought to Europe from Turkey in the 16th century. The true lilies (genus *Lilium*) include the white Easter lily and the colourful tiger lily. The few food plants in this family include asparagus and the genus *Allium*, which includes onion, leek and garlic.

The Amaryllidaceae are mainly tropical plants, but the best-known garden plants are the daffodils and narcissi, bred from the wild species of *Narcissus*, which grow mostly in south-west Europe and northern Africa. These two families and related ones contain plants with large succulent (fleshy) leaves, including the African *Aloe* and the American *Yucca*, one species of which, the Joshua tree, grows as high as a house. An American genus, *Agave*, contains the 'century plant' which pro-duces a huge number of flowers and then dies. It grows for many years before flowering, but not necessarily a century.

The iris family, Iridaceae, like all other monocots, has its flower

Three species of monocots: sweet flag, cuckoo pint and the familiar Easter lily

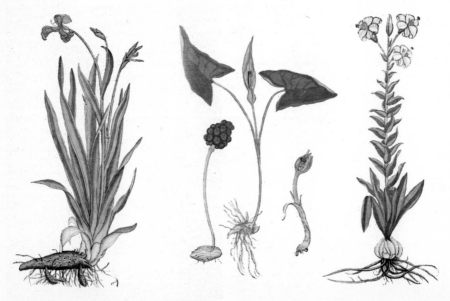

parts arranged in threes, but the iris itself is unusual because its stigmas are carried on styles that look like petals, not stalks. There are many cultivated varieties of the iris family, such as the yellow flag of Europe (the fleur de lys of heraldry) with thick rhizomes, which grow wild in wet places but other members of the Iridaceae family have corms or bulbs.

The gladiolus and crocus are related to the iris. The stigmas of *Crocus sativus*, the saffron crocus, are made into saffron, a yellow powder used to colour and flavour food. This plant is different from the meadow saffron (autumn crocus), which is not a crocus, but a *Colchicum* (Liliaceae) and is very poisonous.

The Araceae is a strange family, with the European wild arum (cuckoo pint) as a typical example. Although it is a monocot, it has net-veined leaves and a mass of tiny flowers on a spike enveloped by a pale green sheath (spathe). The flowers are pollinated by flies and then produce poisonous red berries. Other plants in this family are the calla or arum lily and the sweet flag.

Onions in flower. The white or pale blue flowers are very small and cluster in globular heads

The grass family, Gramineae, is very important to man because it includes both pasture grasses and cereals. Wheat has been cultivated for 10 000 years, and rice, maize (corn), oats, millet and barley are all widely cultivated.

There are over 10 000 species, ranging in size from dwarf mountain grasses to South American grasses twice as tall as a man. Some Asian bamboos are taller than many trees. They all have straight, spear-shaped leaves and often hollow stems with solid joints (nodes). Grass flowers are usually inconspicuous: they are pollinated by wind rather than insects. Lawn and pasture grasses have short or creeping stems which are left undamaged by cutting or chewing, so that when the leaves are mown or eaten by animals they can re-grow from the base of the stem. The scaly wind-pollinated flowers are produced in clusters, each with tiny, colourless petals, but with large stigmas and stamens. The enormous quantity of pollen which they produce is the major cause of hay-fever.

Rice originated as a swamp plant in the monsoon regions; it has been grown in China for at least 5000 years. Today much rice is grown

Left: *papyrus growing wild.* Right: *giant bamboo trees in Ceylon provide both food and shelter*

on drier ground, especially in America and Europe.

Two other Asian grasses are very important: sugar cane, which is now grown mainly in the New World where it was introduced by Columbus, and bamboo. The latter is used for making houses, furniture and poles, and the young shoots are often used as food.

Marsh plants include reeds, which are grasses, and sedges and rushes, which both belong to different families. Sedges have solid three-cornered stems, and rushes have hollow or loosely-filled ones. All three marsh plants are used for thatching, matting and basket-work; one sedge, *Cyperus papyrus*, growing in the River Nile, was used by the ancient Egyptians for making writing paper (papyrus).

Many floating and submerged plants are monocots. The duck-weeds produce very small flowers, and can cover the surface of a pond like a carpet. Submerged monocots include the Canadian pondweed, *Elodea*, and eel grass, *Zostera*. Canadian pondweed choked canals and waterways when it was introduced into Europe in 1842. Eel grass, one of the few flowering plants which can live completely submerged in salt water, is even pollinated under water.

Top: *jointed rush thrives by river banks*. Below: *great duckweed grows in water*. Right: *sedge*

Orchids are found throughout the world, but the most luxuriant grow in the rain forests of south-east Asia and South America. Many grow on the branches of trees. They are not parasites – they merely hold on by means of their special hanging spongy roots. In this way they obtain enough light through the tree tops and also the salts they need from decaying leaves and water from rain.

Orchid seeds are produced in huge numbers but they are small and do not contain enough food for germination. They need to make contact with a fungus which supplies the orchid with food until it can support itself. The fungus then continues to live inside the roots of the orchid and obtains food from it. This mutual help is called symbiosis. Orchid flowers have strange modifications for pollination by insects. A large lump of pollen containing many grains is brushed onto the insect; the lump must be in a particular place on the insect's body so that it can be brushed onto the stigma of another orchid. The flower shape is a perfect fit so that only a special type of insect can enter it – others are excluded. The flowers are very elaborate, and each

Left: *coconut palms, Ceylon.* Right: *the flower-bearing stem of the banana tree has a downward droop. Brightly-coloured male flowers grow at the tip of the stem, while bananas ripen higher up*

may resemble its own pollinating insect, like the European soil-growing bee orchid.

Palms, which grow throughout the tropics, have a single stem, often very tall, with a head of feathery or fan-like leaves at the top. The trunk of a palm does not contain true wood, typical of dicots, but a fibrous tissue. Usually palms have separate male and female flowers in multiflowered spikes. Many species are very useful to man, providing such foods as dates, sago and coconuts, as well as oil, timber and fibres for mats and thatch. Wine is made from the fermented sap of some palms, while others are cultivated for their wax.

The banana tree, family Musaceae, is not a tree at all. The trunk is a cylinder made up of the overlapping bases of the broad leaves. Cultivated bananas have soft pulpy flesh but no seeds: the plant is a sterile hybrid and banana plantations are renewed by planting outgrowths of the old stem. There are two kinds of bananas, one with yellow skins which is eaten ripe and raw as a fruit, and one with green skins which is eaten cooked as a vegetable.

Top: *bee orchids resemble the bees which pollinate them.* Below: *wild orchid of Malaya.* Right: *lady orchids, much smaller than the tropical varieties, grow wild in Europe and western Asia*

All living things must have food to grow. They burn the food to obtain the energy needed for all their activities. In plants the name nutrition is given to all processes involved in obtaining food. The simplest food substances used by animals and plants are chemical compounds such as the sugar glucose. Green plants have a type of nutrition called photosynthesis in which glucose is built up from carbon dioxide and water. Energy from sunlight together with the presence of the green pigment chlorophyll is needed for this process.

In plants other than algae, most photosynthesis occurs in leaves having a flat thin shape which is suitable for trapping the maximum amount of sunlight. The chemical reactions of photosynthesis take place within the plant cells in small capsules (chloroplasts) which contain chlorophyll and a number of enzymes. The job of the chlorophyll is to trap the light energy and convert it to chemical energy. The materials which are necessary for photosynthesis to take place reach the chloroplasts in different ways.

Carbon dioxide from the atmosphere enters the leaf through the stomata (singular: stoma), the small holes in the outer layer of cells

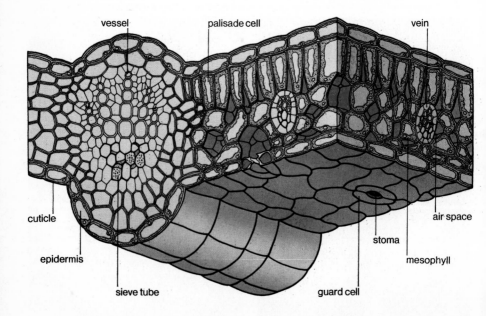

(epidermis). Each stoma is surrounded by two sausage-shaped guard cells. These cells can change their shape and so control the opening and closing of the stomata. The stomata are mainly on the underside of dicot leaves and therefore the carbon dioxide enters from the bottom. It passes up into the next layer, the spongy mesophyll, which is made of irregular cells with large spaces between them. It then passes up further into the palisade layer of cells, which are much closer together and contain many chloroplasts. It is at this level that photosynthesis takes place in the leaf. Above this is the upper epidermis, another single layer of cells. So the leaf looked at in section is like a sandwich: the outer layers are formed of epidermis with the palisade layer and spongy mesophyll inside. The epidermis has a waxy outer coat (cuticle) so that water and gases cannot enter or escape. This is why stomata, the small holes in the epidermis, are necessary.

Water enters the plant through its roots and passes up to the leaf through the xylem, a set of tubes (vessels) that carry water throughout the plant and can be seen in the leaves as veins. Food synthesized in the leaves is carried to the rest of the plant through sieve tubes.

Magnified photograph of a leaf cross-section. Compare with the diagram on the facing page

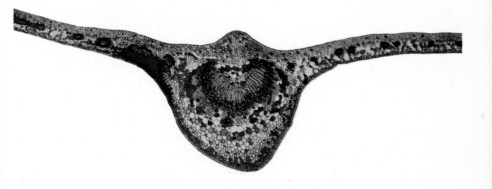

The plant kingdom
40   photosynthesis: using the sun's energy

All green plants obtain food by photosynthesis – a process in which carbon dioxide and water are chemically combined to form the simple sugar glucose. Oxygen is formed as a waste product and is the source of most of the oxygen in the air we breathe. The reaction occurs in chloroplasts, the small rounded structures within the cells of the plant that contain chlorophyll.

We usually think of light as waves of radiation, but it can also seem like a stream of separate particles called photons. When these photons hit the chloroplasts the light energy is converted into chemical energy by the chlorophyll and oxygen is released from water. This stage of photosynthesis is called the light reaction. The chemical energy then brings about the reaction between carbon dioxide and the hydrogen from the water, called the dark reaction as it does not require the presence of light to take place. The reactions occurring in photosynthesis can be expressed as a chemical equation:

$$\text{carbon dioxide} + \text{water} \xrightarrow[\text{sunlight}]{\text{chlorophyll}} \text{glucose} + \text{oxygen}$$
$$6CO_2 \qquad + 6H_2O \qquad\qquad C_6H_{12}O_6 + 6O_2$$

In higher plants carbon dioxide is obtained from the air and water

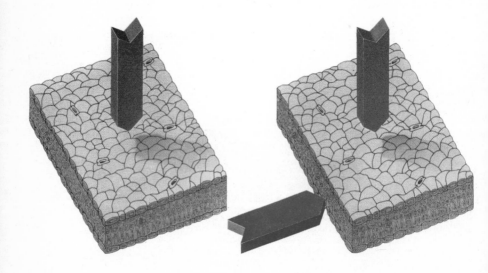

section of leaf

comes from the soil through the roots. Oxygen produced during the reaction is returned to the air. The glucose is converted by plants into other chemical compounds such as starch, cellulose and fats. It can combine with nitrates from the soil to form amino acids and proteins.

It is important to understand that green plants are the only organisms that can make food from simple molecules such as carbon dioxide and water. Some animals eat these plants; others feed on the plant-eating animals. Thus, photosynthesis is the basis for all life as all animals must depend eventually on plants for their food. Photosynthesis is also important as it helps to keep a correct balance between the oxygen and carbon dioxide in the air. During respiration oxygen is removed from the air by plants and animals. The oxygen would soon all be used up if it were not returned to the air by plants during photosynthesis. Carbon dioxide is returned to the air during respiration but it would soon poison all life if it were not removed during photosynthesis. The amount of carbon dioxide and oxygen in air remains constant (oxygen at 20% and carbon dioxide at $0\cdot3\%$) due to a balance between the rates of photosynthesis and respiration.

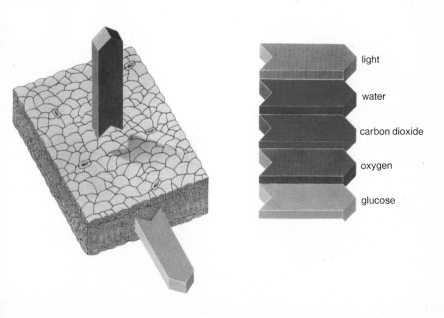

light

water

carbon dioxide

oxygen

glucose

The plant kingdom
41   respiration: the release of energy

Respiration is the oxidation (breakdown, using oxygen) of food inside the living cells of animals and plants to release energy. The energy is used for many chemical processes in the cells. Respiration is like photosynthesis in reverse: oxygen is taken into the cell and breaks down glucose into water and carbon dioxide, releasing chemical energy. The reaction is expressed by a chemical equation:

$$\text{glucose} + \text{oxygen} \longrightarrow \text{carbon dioxide} + \text{water} + \text{energy}$$
$$C_6H_{12}O_6 + 6O_2 \qquad 6CO_2 + 6H_2O + \text{energy}$$

Photosynthesis uses the sun's energy to make food; respiration breaks down and releases the energy. In growing plants there is a continual supply of glucose, formed during photosynthesis; in seeds, bulbs and other resting organs, stored food, often in the form of starch, must first be converted into glucose, which is then oxidized to release energy. The chemical energy released during respiration in plants is eventually converted into different types of energy needed for different jobs in the cell. For instance, some is converted into mechanical energy for pushing the roots through the soil, and some is used in making plant tissues such as cellulose and wood. Germinat-

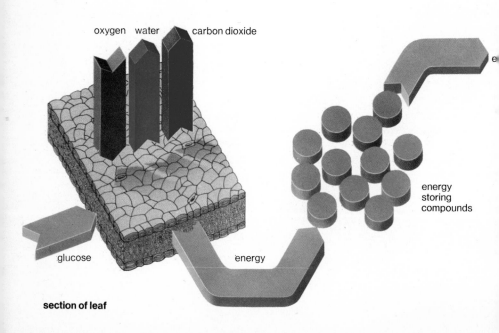

oxygen   water   carbon dioxide

energy
storing
compounds

glucose   energy

section of leaf

ing seeds need a lot of energy for growth into seedlings and therefore have a very high respiration rate. However, the chemical energy from respiration cannot be used directly by the cell. It is first stored in the compound ATP (adenosine triphosphate). When energy is required the ATP is converted to ADP (adenosine diphosphate) and the energy is released. These reactions can be summarized as:

when energy is stored: ADP + phosphate + energy $\longrightarrow$ ATP

when energy is required: ATP $\longrightarrow$ ADP + phosphate + energy

During respiration both plants and animals take in oxygen and give off carbon dioxide and water vapour. In animals the term 'breathing' is used for the movements accompanying the exchange of oxygen and carbon dioxide, such as the filling and emptying of the lungs. The exchange of gases in plants occurs by diffusion between the surface of the plant and the atmosphere.

Plants do not always use oxygen in respiration. Some plants, such as yeast, can still respire if there is no oxygen present, but they only obtain enough energy to stay alive without growth, and instead of water they produce alcohol.

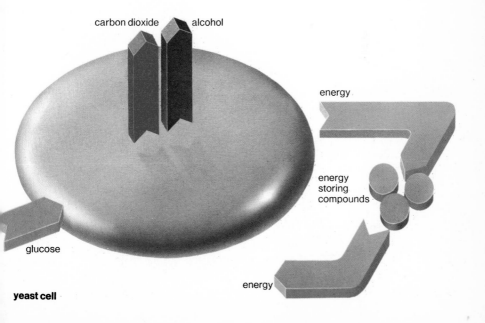

carbon dioxide   alcohol

energy

energy
storing
compounds

glucose

energy

yeast cell

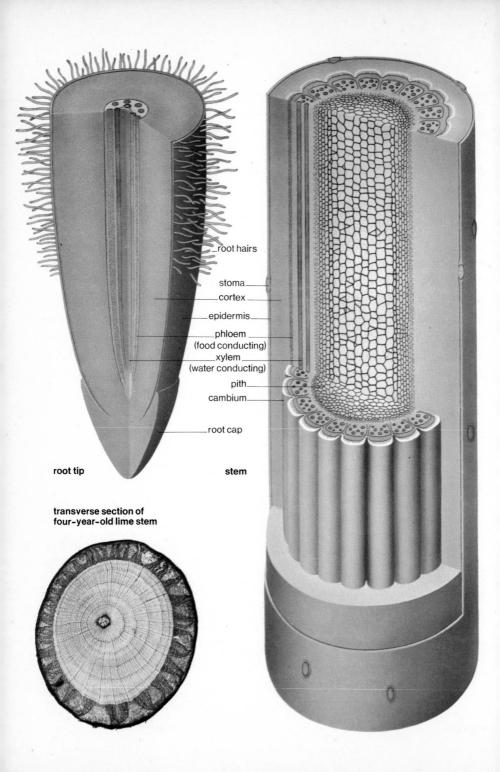

root hairs

stoma
cortex
epidermis
phloem
(food conducting)
xylem
(water conducting)
pith
cambium

root cap

root tip

stem

**transverse section of
four-year-old lime stem**

Most plants require a transport system for carrying materials. Instead of a continuous circulatory system such as that of the human body, plants have two sets of pipes. Each consists of long tubular cells; one set carries materials upwards and the other downwards. In almost all land plants the 'up' transport system consists of xylem vessels, which are wooden pipes that carry water from the soil through the roots and up the stem. The 'down' system is made of long thin-walled tubes called phloem cells. The phloem system transports food from the leaves where it is made to parts of the plants where it is used or stored. It can transport up to the buds and growing points as well as down.

Water makes up over three-quarters of most living things and is constantly used up in cell processes. Small plants are kept upright by being distended with water just as a motor car tyre is firm when filled with air. A tyre when it is flat cannot support the car, and a plant short of water on a hot day wilts. Most of the water taken up by a plant is lost by evaporation from the leaves, a process called transpiration. A maize plant loses about two litres a day because when the stomata are open to exchange gases during photosynthesis, water vapour is lost very rapidly. Transpiration actually helps to draw water up the plant in a way similar to sucking liquid up a straw.

Water and dissolved minerals are collected by thousands of minute hairs located just behind the growing tip of each root branch. It has been calculated that a rye plant one month old and 50 cm tall has 15 000 000 000 root hairs with a total length of over 8000 km.

Roots are very strong and supple because the woody xylem vessels are concentrated at the centre. In stems, however, xylem and phloem are arranged in bundles, either in an outer ring or scattered throughout the stem like reinforcing rods. This gives the stem strength to bear the weight of branches and to withstand the strain of wind. The mass of any living tree is made up of dead xylem vessels (wood); new, living wood cells are produced each year. More and bigger xylem vessels are produced in spring, forming spring wood. Smaller xylem vessels are produced in summer. The difference between the light-coloured spring wood and the dark wood of the previous summer shows up as an annual ring by which it is possible to tell the age of a tree.

Green plants – those with chlorophyll – make their food with the help of sunlight energy. But some plants do not possess chlorophyll and must obtain their food ready-made, in a way similar to animals. Most fungi and many bacteria obtain their food by dissolving or digesting dead plant and animal remains or other organic material. This action is what is really meant by decaying. Plants acting in this way are saprophytes and do an important job in breaking down industrial waste, sewage, excrement and dead plants and animals.

There are no true flowering plant saprophytes. *Neottia*, called the bird's nest orchid because its tangle of roots and underground stems looks like a bird's nest, is sometimes regarded as a saprophyte. It obtains food with the aid of a fungus in its roots. The fungus is saprophytic on dead matter in the soil and the orchid is parasitic on the fungus. The only part above ground level is the leafless flower stalk.

One of the best-known parasites (plants which get their food from a living host) is mistletoe, which grows mainly on apple and black poplar trees. It contains chlorophyll and makes its own food. But mistletoe draws its water and minerals from its host through suckers that penetrate the tree and join with the host's water-conducting system. Therefore it is only a partial parasite. Dodder is an almost leafless, completely parasitic plant, consisting of a mass of threads twisted around the stem of its host, which is often gorse or heather. Its suckers enter the host's transport system to obtain its nourishment. Another complete parasite is the giant Malayan stinking lily, *Rafflesia*, which, apart from its huge flower, consists only of absorbing threads buried in the roots of its host vine.

Some plants which live in poor soils gain extra minerals by killing and digesting insects. These are known as carnivorous plants. Sundew, *Drosera*, found in boggy, peaty areas in Britain and elsewhere, catches insects on sticky 'tentacles'. The Venus' fly-trap, *Dionaea*, from the Carolinas in the United States has hinged leaves which snap shut, while many of the pitcher plants, such as the Asiatic *Nepenthes*, drown insects in digestive fluids in their pitchers.

Contact with the trigger hairs on the inner surface of the Venus' fly-trap leaves can snare wandering insects. This unfortunate frog came too close to the marsh-dwelling plant

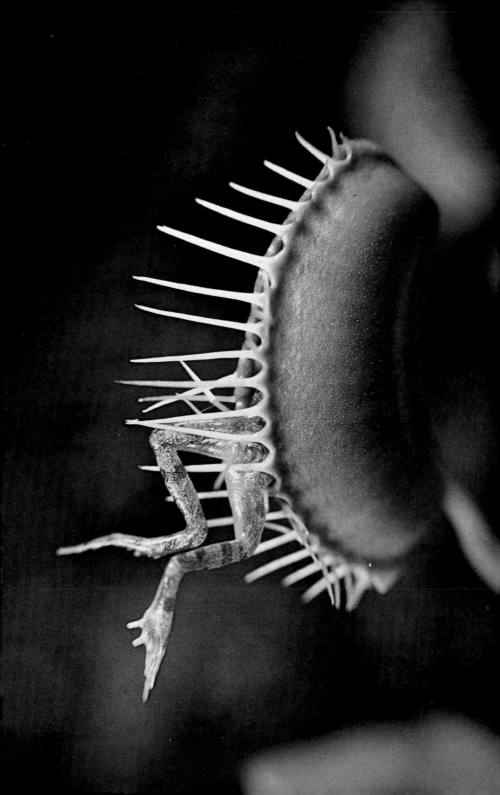

Plants exist everywhere, and their roots, stems and leaves are often modified so that they can adapt to special environmental conditions.

Roots are not always underground organs: some plants have aerial roots. Certain Malayan orchids, growing high up on jungle trees, have practically no stems, and their leaves are no more than brownish scales; all photosynthesis takes place in the green aerial roots. Ivy, *Hedera helix*, anchors itself to a wall or other support by aerial roots. The most striking of all aerial roots are the dense prop roots of the mangrove of tropical America and banyan trees of India, both of which grow in swamps.

Plants that live in particularly dry places are known as xerophytes. They are modified in various ways to reduce the loss of water. Succulent plants, such as cacti and houseleeks, store water in thick stems. Others have leaves which are thick-skinned, hairy, or reduced to spines to cut down evaporation of water. Some of the tree-sized cacti of the North American deserts store enough water in their stems to last for many years.

Left: *grapevine tendrils, formed from modified stems, help the plant to climb and train over trellises.* Right: *rose thorns are also aids to climbing because they cling to walls or other plants*

Leaves, stems and roots may be modified to act as climbing organs. The pea has tendrils formed from leaves, clematis has them formed from leaf stalks, while the grapevine and passion flower have them formed from stems. Many plants climb by twining: hop and honeysuckle twist clockwise; runner bean and bindweed twist anti-clockwise. Roses and blackberries climb with the help of thorns, which are outgrowths of the stem.

Lianas are woody climbers and are rooted in the ground, no matter how high they climb. Ivy and honeysuckle are found in the temperate zones and lianas of the tropical rain forests have stems as wide as 30 cm in diameter.

Some of the most beautiful flowering plant epiphytes are orchids. Epiphytes are plants that grow high up on their supporting trees and do not root in the ground. Many different plants have an epiphyte habit including lichens, mosses, ferns and angiosperms. One of the strangest is Spanish moss, a flowering plant that hangs in dense grey masses from trees in the rain forests of Central America.

The aerial prop roots of the banyan tree make it look wider than it really is. In India, where the tree is sacred because of its association with Buddha, the roots are very carefully tended

A flower cannot produce seeds until it is pollinated and its ovules fertilized. Pollination is the transfer of pollen from the male parts (stamens) to the female parts (stigmas) of a flower. If pollen is carried to the stigma of the same flower, it is called self-pollination. More often, pollen is carried to a flower of another plant of the same species. This is called cross-pollination. The advantage of cross-pollination is that characteristics from two plants combine in the seeds, producing great variety in the offspring that develop from them. However, many plants, such as some varieties of apple and pear, are self-sterile, which means that self-pollination cannot take place.

Pollen is usually carried from one plant to another by the wind or by small animals, mainly insects. Wind-pollinated flowers are usually inconspicuous and small, with tiny scaly petals but large feathery stigmas to catch the pollen. Grasses and most of the catkin trees are wind-pollinated, and their pollen is produced in vast quantities, causing hay-fever in some people.

Flowers pollinated by insects and other small animals have large brightly-coloured petals or scent which attracts animals. But in order

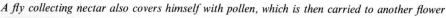

A fly collecting nectar also covers himself with pollen, which is then carried to another flower

to make the visit worthwhile to insects and animals, nectar is produced at the base of the petals. Nectar is a sugary liquid used by many insects for food, and collected and converted to honey by bees. The pollen sticks to the insect while it is sucking the nectar, and some is deposited on the stigma of the next flower it visits. Bees are especially attracted to sweetly-scented blue, yellow and purple flowers, and moths to white flowers with a very strong night-time scent, such as the tobacco plant. Flies and midges favour plants which, to us, have an unpleasant smell. Larger animals such as hummingbirds and bats pollinate large tropical flowers.

After a flower has been pollinated, fertilization can take place. When a pollen grain lands on a suitable ripe stigma it quickly grows a tube down through the style and into the ovary to the ovule. A male cell then passes through the pollen tube and fertilizes the egg cell. After fertilization has taken place the ovule develops into a seed, which consists of an embryo plant together with a food store and a protecting coat. At the same time the ovary develops into a fruit as its wall either dries out or becomes fleshy and juicy.

Catkins on a hazel tree are really clusters of tiny flowers

A fruit is the ripened ovary of a flower, enclosing seeds. Many vegetables, such as tomatoes, cucumbers and runner beans, are really fruits, and the cultivated banana is a fruit in which the seeds do not develop properly. As plants stay rooted in one place, their fruits are almost always adapted in some way for dispersing, or spreading, the seeds to ensure that new plants do not take up valuable soil space that is needed by the parents. Some fruits, such as the pods of gorse and pea, burst open when they are dry and shoot out the seeds. A ripe poppy head scatters seeds through small holes in the seed case when it is shaken by the wind. Many wind-dispersed fruits have wings, such as the ash and sycamore. The fruits of thistle and dandelion have a parachute of hairs, and wild clematis, or old man's beard, have a hairy style attached to their fruits. The cotton plant has a hard dry fruit which splits open to release a mass of hairy seeds. Fruits and seeds are often carried a long way by the wind.

Many fruits are dispersed by animals. Blackberries, cherries, plums and apples are eaten by animals, and the seeds are either discarded or

Left: *the bright colour of rose hips attracts birds, which eat the fruit and thus spread the seeds.*
Right: *sycamore seeds have large wings so that they can be scattered by the wind*

passed unchanged through their digestive systems. Birds disperse fruits such as the rose hip in the same way. Animal-dispersed fruits are often brightly coloured yellow or red to attract the animal, and have a sweet taste which encourages the animal to return. Nuts are another type of tasty fruit that is dispersed by squirrels and similar animals. Coconut seedlings are found growing in the sand of many Pacific islands, which suggests that some fruits are carried by water.

A seed will remain dormant until conditions are suitable for its germination. The first part to emerge is the young root (radicle) which grows down into the soil, then the young shoot (plumule) begins to grow upwards, drawing on the food store in the seed until the first leaves are formed. The shoot always grows upwards, and the roots downwards even if the seed is planted upside down, because the plumule grows away from the force of gravity, and the radicle grows in the same direction as gravity. The shoot also responds to light and grows towards it as soon as it reaches the surface. This is why a plant placed on a window sill soon bends towards the light.

In old man's beard, a type of wild clematis, each seed is attached to a long feathery style so that it can easily soar and be dispersed

Many flowering plants can reproduce without seeds. This is called vegetative reproduction. It is often used by gardeners because the plants develop faster, are less delicate and thus not easily damaged while growing. The new plant has exactly the same qualities as its parent. There are many forms of vegetative reproduction, such as rhizomes, bulbs, corms, cuttings, runners and tubers. Rhizomes, such as couch grass and cultivated irises, are creeping underground stems from which many shoots grow upwards, forming new plants.

Another common way for plants to multiply is from bulbs and corms. A bulb is a shoot surrounded by swollen fleshy leaves which serve as a food store for the shoot during the early stages of its growth. The fleshy leaves are surrounded by brown papery scale leaves. In the flowering season the food in the fleshy leaves is used to supply the growing plant and the leaves wither. After flowering more food is formed by the aerial parts of the plant, which is returned to the base of the stem to form a new bulb. The bulb remains dormant during the winter and produces a new plant the following spring. Around the parent bulb grow a number of little bulbs which split off to form new

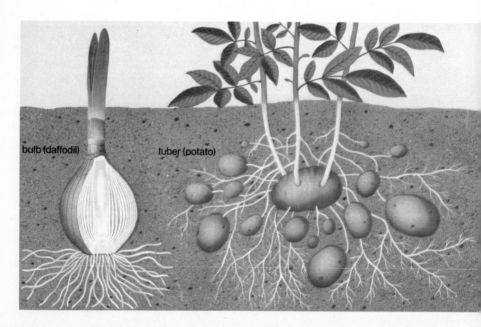

bulb (daffodil)   tuber (potato)

plants. Onions, daffodils and tulips are grown from bulbs. Daffodils grow wild in Europe and from them gardeners have bred the modern large varieties. The tulip originally came from Persia where it had been cultivated for hundreds of years, and was brought to Holland in the 16th century. Its name means 'turban'. It created a sensation in Holland, but some Dutch growers were ruined by a 'tulip mania' because the imported bulbs were very expensive. Corms, such as crocus and gladiolus, look like bulbs but store food in a swollen stem.

There are many other forms of vegetative reproduction. Cuttings are cut pieces of stem which take root when planted. In some begonias and other species a cut leaf will take root and grow into a new plant. Strawberries send out runners – thin stems creeping along the surface of the soil, at the end of which a new plant grows. Potatoes multiply by their tubers, which are really underground stems swollen with starch, and each potato 'eye' can grow into a new plant. All potatoes of any one variety originally came from the same plant. The cultivated banana can only be grown from shoots that come from older plants because its seeds never fully develop.

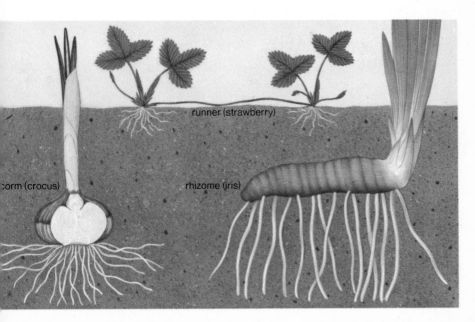

runner (strawberry)

corm (crocus)   rhizome (iris)

The plant kingdom
48 farmers and the soil

Agriculture involves tilling the ground, planting crops, tending and harvesting them, and planting again. It began thousands of years ago in the fertile river valleys of the Tigris, Euphrates and Nile, where annual floods spread rich mud over the crop fields. Beyond the fertile valleys farmers had either to irrigate dry land or clear a space at the edge of the forests before grain could be planted. This soil was full of decaying organic material (humus) which was used up by the crop plants. Once the humus-rich soil was exhausted, more trees had to be cut down and a new space cleared. If the farmers allowed trees to grow when they left an area, their descendants could return many years later to find that the soil had been refertilized by decaying tree vegetation. But if the abandoned farmland was continually grazed by sheep and goats, and the trees had not been allowed to grow, the soil was exposed to wind and rain and often turned to dust. This is how vast areas of the Old World, such as the Sahara, became desert. Many hundreds of years later, in the 19th and 20th centuries, the dust bowls were created in the same way in the middle-western areas of the United States.

Wall paintings, found in an Egyptian tomb, show how ancient farmers tended their crops

Plants can be cultivated continuously on the same piece of land if the soil is fertilized. In the Middle Ages soil was kept fertile by growing crops on a three-field rotation. Each field in turn yielded crops for two years and lay fallow for the third. In the 18th century farmers used a four-stage plan: clover in the first year, root crops in the second and cereals in the third and fourth. Clover helped to refertilize the soil with nitrates. The bacteria in their root nodules could manufacture their own nitrates, and when the plants decayed, the nitrates in the roots were released to enrich the soil. Thus the mineral content of the soil was kept in balance. In the 19th century farmers regularly dug in organic fertilizers such as farmyard manure, fish and bone manure and sea-bird droppings (guano) from the coastal islands of Peru, Africa, Chile and the West Indies. Modern farmers use inorganic fertilizers: mostly large quantities of nitrates, potash and superphosphates, and traces of iron, manganese, copper, zinc and boron.

Today yields are increased by reducing the number of insects that eat the crops. Aircraft with crop-dusting equipment are often used to spray insecticides over large areas of farmland.

The labour-saving devices used by today's farmers include crop-dusting from the air

The plant kingdom
49   cereals: the world's basic food

Cereals are plants of the grass family whose seeds (grain) are used for food. A cereal grain consists of a small embryo (germ) that will grow into a seedling, and a supply of food (endosperm) to nourish it. The food is mainly starch, although the grain also contains other foods such as proteins and vitamins.

Wheat, the oldest and most widely cultivated cereal, is grown in most non-tropical areas of the world. Its grains are ground (milled) to make flour. If the endosperm alone is milled, white flour is produced; if the whole grain is used, brown or wholemeal flour is obtained. This is more nourishing than white flour because it contains proteins and vitamins as well as starch. Wheat was probably first cultivated in Neolithic times, about 8000 years ago, in Asia Minor. For a long time only a few species were cultivated, such as emmer and spelt, from which modern wheat developed. More recently selective breeding has produced varieties to suit most climates and to resist most diseases. The breeding process still continues, and the yield per hectare has risen considerably in the last 20 years.

Rice is the staple food of over half the world's population, par-

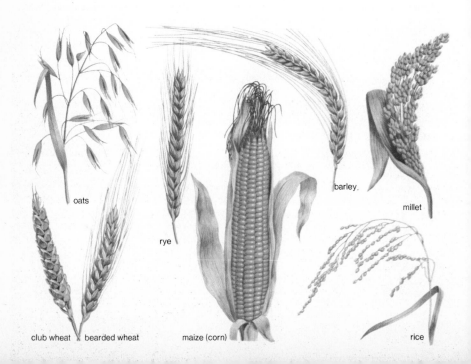

oats

rye

barley

millet

club wheat | bearded wheat   maize (corn)   rice

ticularly in southern and eastern Asia. It is mostly grown as a swamp crop and the seed is planted either directly in the flooded paddy fields or in well-fertilized soggy nursery beds from which the 20 cm tall seedlings are later transplanted into the fields.

Barley is a cereal grass which is widely grown for animal and human food and to make beer and whisky.

Rye is coarser than wheat and is grown only in soils too poor for wheat, such as parts of northern and eastern Europe and Russia. It is used to make black bread and also for cattle food. Oats, which can be grown in moist climates with short summers, have more loosely-arranged ears and are used to make oatmeal for porridge and cakes, and as animal food.

Maize (corn) originated in the Americas and is a much bigger plant than other cereals. It is also bushier, sometimes grows man-high and is sensitive to frost. Maize is widely used for animal food but some is used for human consumption, as for example in cornflakes, and the young heads provide sweet corn. Another edible grain, mainly a tropical cereal, is millet (sorghum).

In Indonesia rice is planted in terraces, cut into the hillsides, which are carefully flooded and tended

Sugar cane, *Saccharum officinarum*, is a giant grass, three to six metres tall. It was taken to the New World by Columbus and today it is a main crop in the West Indies and Brazil. It needs a hot moist climate and a rich, well-drained soil. It is grown from stem cuttings, not seeds, and sugar is obtained from the stem (cane). The cane is cut either by machine or with a machete, a longish hand-held blade. The sap is crushed out of the cane, dissolved in boiling water and finally refined to white sugar. The yield per hectare is about 12·5 tons.

In more temperate climates sugar is obtained from the roots of the sugar-beet, *Beta vulgaris*. This plant and the beetroot have been developed and cultivated by man from the same wild plant which grows on European sea-shores. The long, white, pointed roots of the sugar-beet are sliced and the sugar dissolved in boiling water. The beet tops are sold as cattle food. The yield per hectare is about five tons. Sugar-beet is an important European crop and produces about a third of the world's supply of sugar. There are many other root crops – plants whose roots or underground stems store starch or

This engraving by a 16th century German artist shows slaves at work on a sugar plantation on the West Indian island of Hispaniola (now known as Haiti and the Dominican Republic)

sugar – which can be used as food by man. The potato, which contains starch, comes from the Andes Mountain areas of South America, and was brought to Europe about four hundred years ago. The wild plant produces small, tasteless potatoes, but cultivated varieties, of which there are now very many, have been bred to yield much larger, tastier potatoes. The edible part is, in fact, not a root, but a swollen underground tuber, which grows well in moist soils in cool, damp conditions. The yield of potatoes, in weight per hectare, is higher than cereals, but much of it is water, so it takes about four kilograms of potatoes to equal the food value of one of wheat.

The tropical sweet potato, quite different from the common potato, is a climbing plant related to the bindweed, *Convolvulus*. Its starch-filled tubers can be boiled and either mashed or ground into flour, and its leaves are used for animal fodder. Other tropical root crops include arrowroot, yams and cassava (manioc), which is the source of tapioca. Some root crops of the temperate regions are turnip and swede, which are crucifers, carrot and parsnip, which are umbellifers, and onion, shallot and leek, members of the lily family.

coffee crystals   light brown   molasses   granulated   golden syrup   castor   barbados   dark brown

The fleshy fruits grown by man have all been obtained by selective breeding from wild species. The parent species of all cultivated apples is *Malus pumila*, the wild crab apple of Europe, Asia and North America, a thorny plant with small acid-tasting green apples. Today there are more than 3000 named varieties, many of which are grown as dwarf forms rather than large trees. The plants are not grown from seeds but by cuttings from high-quality plants grafted onto low-growing, less valuable root stocks. The graft plant bears apples after only one or two years. Stone fruits are those with a large stone in the centre which surrounds the seed, and all of them belong to the genus *Prunus*. Their stones often have a slight flavour of almond, which also belongs to this genus. Plums, including damsons, sloes and gages, grow wild throughout the temperate zone; peaches and apricots are native to China, but the cultivated form of both types is grown all over the world. Stone fruits are usually grown like apples, by grafting.

Top: *figs are tropical and subtropical fruit but lemons* (below) *are subtropical only*

In the bush fruits blackberry and raspberry, each fruit is really a collection of tiny fruits joined together. The loganberry is a man-made hybrid of the blackberry and raspberry. Cultivated strawberry varieties have been derived from American species, although there are several kinds of European wild strawberry.

All fruits described so far belong to the family Rosaceae but some plants not in this family also produce juicy fruits. Many hillside shrubs such as bilberries and cranberries produce delicious fruits. Citrus fruits such as oranges, lemons and tangerines originated in China and south-east Asia, while grapefruit is believed to be a native of the West Indies. Other tropical fruits include the Old World fig, pomegranate and mango and the New World pineapple, passion fruit and guava. Many fruits are cultivated far beyond their native regions; for example, the biggest banana-growing country today is Ecuador, closely followed by the West Indies, parts of Africa and Australia, but bananas are also grown in India and south-east Asia.

Blackthorns (top) *and raspberries* (below) *are both grown in temperate climates*

Early in the development of agriculture men discovered how to make alcoholic drinks from grapes and corn. The ancient Egyptians drank both wine and beer, and the Greeks carried on a lively trade in wine throughout the Mediterranean. The vines of grapes are all of a single species, *Vitis vinifera,* although there are hundreds of varieties adapted to different soils and climates.

Wine is the fermented juice of fresh grapes. The juice of the wine grape contains sugar, and growths of yeast form on the outside of the grape skins. In wine-making, the grapes are crushed in a wine press and the yeast converts the sugar to alcohol, when there is no air present, by a process called fermentation. Red wine is made from dark grapes, and white wine from white grapes or from dark grapes whose skins have been removed from the wine press at an early stage. The most famous wine-growing countries are France, Germany and Italy. Wine was made in England in the Middle Ages, but the climate is not really suitable for grapevines. Wines must be drunk quickly once they are opened, otherwise bacteria will use the air to convert the alcohol to vinegar. The bacteria are killed by a higher alcohol

Grapes growing in a vineyard in Crete

content than is found in wine and that is why sherry and port, the specialties of Spain and Portugal, are fortified by the addition of spirits to make them last longer.

Beer is made from sprouting barley grains (malt) which is fermented with yeast to produce alcohol; hops are added for flavour. Ale, the most common drink in England in the Middle Ages, was also made from barley, but without hops; the ale of today is merely a type of beer. In Japan beer is made from rice.

Spirits have a higher alcoholic content than beer and wine and are made by distillation from a base of grain or some other vegetable. Gin and vodka can be distilled from a variety of ingredients, including potatoes; gin is flavoured with juniper berries. Scotch whisky is obtained from a base of fermented barley, and brandy from the distillation of wine. Rum is derived from sugar cane by fermentation of molasses, a by-product in refining sugar. Cider is made from apples. South American Indians make alcoholic drinks from cactus leaves and the shoots of certain palm trees.

Treading the grapes to extract their juice was one of the earliest methods of wine-making. The men shown here in Madeira are still using this method to make their famous wine

The plant kingdom
53   tea, coffee and cocoa

Tea, coffee and cocoa, the main non-alcoholic drinks, are all obtained from trees or shrubs but they were unknown in Europe until the 16th century. The Chinese were probably the first to drink tea, about 4000 years ago, because the tea shrub, *Camellia sinensis*, is native to China. It is now cultivated in many other countries, especially India and Sri Lanka (formerly Ceylon). Tea is made from the young leaves. Green tea, popular in China, is made by drying the leaves quickly; black tea is made from leaves that are pounded and fermented before they are dried. Tea became very popular when it was introduced to England in the 17th century. Today England takes a third of all tea exports and Russia, Australia and Iran are also large importers. Tea contains the mild stimulant theine, a chemical similar to the caffeine in coffee.

Coffee is as popular a drink as tea. It is made from the dried seeds (coffee beans) of the evergreen shrub, *Coffea*. The shrub grows wild in parts of Africa, and Kenya is one of the most important African coffee-producing countries. Coffee was introduced into Arabia in the 15th century and the Arabs, who are forbidden by their religion to

Left: *in a Mayan painting, dated about A.D. 1300 from Yucatan, Mexico, the young god of vegetation holds a bowl containing three cacao pods.* Right: *tea-picking in Nepal*

drink alcohol, became great coffee drinkers. Their merchants sold the beans in Europe in the 17th century and coffee soon rivalled tea in popularity. Later, coffee was found to grow well in South America, especially in Brazil, which became, and remains, the giant among coffee producers. Today the United States consumes more coffee than any other country.

Cocoa is made from the fruit of a small tree, *Theobroma cacao*, which is native to South America and contains the stimulant theobromine. The flowers and fruit grow in a strange way directly from the trunk. The ripe seeds are fermented, dried, and ground to a paste to produce bitter chocolate from which the Aztecs and Mayans made their favourite drink. After the conquest of Mexico, the Spaniards brought the new beverage to Europe, but it was not drunk in large quantities until the 17th century. Eating chocolate and powdered cocoa were not developed until the 19th century. The countries where the tree is grown are Brazil, Ghana and Nigeria. It requires a very special sort of climate, with warm, shady, humid conditions and a heavy rainfall.

Coffee growing on a plantation in Kenya

Drugs are chemicals that affect the functioning of the body. They are used medically to combat disease and some are taken for the pleasant sensations they induce. Until the 19th century most drugs came from plants. In the 17th century American Indians discovered that chewing the bark of the cinchona tree often cured them of malaria, but it was not until 1848 that quinine was isolated from the bark as the substance responsible for the cure, and its chemical structure discovered.

Many drugs are taken by people to change their mood or state of mind. Continued use for this purpose often leads to addiction – a strong desire which cannot be resisted. The continued use of some drugs may bring about a serious or fatal illness because large quantities taken over a long period of time have a poisoning effect on the body. Some of the addictive drugs are also used medicinally, but unless they are prescribed by a doctor their use is usually illegal.

Opium is one of the major addictive drugs. It is derived from the seeds of the opium poppy and has been chewed and smoked in the Orient for thousands of years. It dulls pain and hunger and produces a feeling of well-being. In time the opium addict becomes lazy and

A South American woman chews a coca leaf

weak-willed and if he continues taking the drug, will die prematurely. In the 19th century the opium derivatives morphine and heroin were discovered, and they proved to be better pain-killers than opium itself. In South America the Indians chew the leaves of the coca tree, which produces similar effects to opium. Cocaine extracted from the leaves was found to be the active agent, and this substance was used by dentists as a pain-killer until a safe synthetic form was made.

Opium and its derivatives and cocaine are addictive and dangerous. Another long-known and now illegal drug is hashish, also called Indian hemp, cannabis, marijuana and pot. The word assassin means hashish-eater, and was originally the name of members of a fanatical hashish-eating Muslim sect who murdered people by command of their leader.

Among the many other drugs derived from plants is curare, a poison in which American Indians dipped their arrows. Today it is used medically in small quantities. Digitalis, obtained from the fox-glove, is sometimes used to treat people who are suffering from certain forms of heart disease.

Left: *foxglove*, Digitalis purpurea, *(16th century engraving)*. Right: *centre of an opium poppy*

Most natural rubber is obtained from the latex of the rubber tree, *Hevea brasiliensis*, native to the tropical forests of the Amazon. Latex is a milky fluid similar to the fluid in the stems of dandelions and poppies. It runs out of the tree when a semicircular cut is made on the trunk and is then collected in bowls and treated with chemicals to thicken and harden. Until the end of the 19th century, Brazil supplied the world with all the rubber it needed. But when the production of bicycles and cars increased there was a huge demand for rubber for tyres which the small number of widely-scattered wild trees of the Amazon could not meet. In 1876 an Englishman took some Brazilian rubber trees to the Botanic Gardens at Kew. These provided the seedlings for the vast rubber plantations developed in Malaysia and which now produce a large proportion of the world's supply of natural rubber. The invention of synthetic rubber has reduced the demand, but over 2 000 000 tons are still produced each year.

Many plants, especially tropical species, produce globules of oil or fat which are stored in the fruit or seed. The coconut palm, *Cocos nucifera*, widely distributed in tropical regions, is one of the most

Olive plantations, seen here in Libya, are found in nearly all middle-eastern countries

important: its oil is used for margarine, cooking oils and soap, and the residue for cattle food. Other tropical palms provide oil for paints and industrial purposes. The soya bush is the most valuable legume crop of China and Japan, bearing seeds (soya beans) which are rich in edible oil. Peanuts, also called groundnuts, sunflower seeds and cotton seeds are all important sources of vegetable oils.

Outside the tropics, the most important vegetable oil comes from the olive tree, *Olea europaea*, grown since ancient times in Mediterranean countries. Olive oil, a clear golden-yellow fluid, is pressed from the ripe black olives. The best olive oil is used for cooking, in salad oils and for packing sardines; second-grade oils are used to make soap and in some industrial processes.

Oils for making perfume are obtained from plants such as the rose, lavender and jonquil, which grow in temperate regions, either by distillation or by dissolving out the oil with alcohol. The product is called an essential oil and is very expensive; for example, over one hundred thousand blooms are needed to produce only one litre of the essential oil of roses (attar of roses).

Left: *sunflowers produce valuable vegetable oils.* Right: *tapping latex from a rubber tree*

Plant fibres are mainly used for making cloth. Cotton comes from the plant *Gossypium*. The actual cotton fibres are long hairs that grow out of the seed coat. The seeds are within fruits called bolls, and a ripe boll looks like a mass of cotton wool, about the size of a golf ball. The long cotton fibres (lint) are removed from the seed coat by a process called ginning, and spun into cotton thread which can be woven into cloth. Most cotton is grown in the United States, South America, Russia, Egypt and the Sudan. It is known to have been cultivated 5000 years ago in Peru and parts of Asia.

Linen goods have been used since prehistoric times. Flax, the fibre used to make linen, comes from the stems of a plant, *Linum usitatissimum*, which is 30 to 100 cm high and has small white or blue flowers. The flax fibres are, in fact, part of the phloem. The fibre is removed from the plant stems by retting, which is soaking them in water so that the rest rots away. The fibres are then spun into thread and woven into linen cloth. Most flax today is grown in Europe.

The fluffy material used to stuff pillows and cushions is called kapok. It comes from the kapok (silk-cotton) tree, *Ceiba pentandra*,

The processing of flax was done by hand for many hundreds of years

which grows from 18 to 30 metres high and bears hundreds of pods, each 15 cm long. The fibres (floss) are inside the pods. When the pods are picked the floss is pulled out by hand, dried, cleaned and exported in bales. Most cultivated kapok comes from Java, although the tree is found throughout the tropics.

Two important fibres used to make coarser cloth and rope are hemp and jute. There are several plants called hemp, such as the true hemp, *Cannabis sativa*, sisal hemp and Manila hemp; in each case the fibre comes from the leaves and is used to make rope, cord and twine. Jute is the phloem from the stem of an Asian plant (*Corchorus*). Like flax, it is obtained by retting, and is used for sacks, hessian cloth, and as a backing for linoleum and carpets. Almost all the world's jute comes from India, Bangladesh and Thailand. Another useful material of plant origin is raffia. This is obtained from the leaves of a Madagascan palm tree, *Raphia ruffia*, and is woven into rush baskets and hats or formed into strips for tying up plants.

Today, man-made fibres such as rayon, nylon, acrylics and many other synthetic products are partly replacing the natural fibres.

Left: *cotton plants bearing ripe cotton bolls.* Right: *unripe pods of a kapok tree*

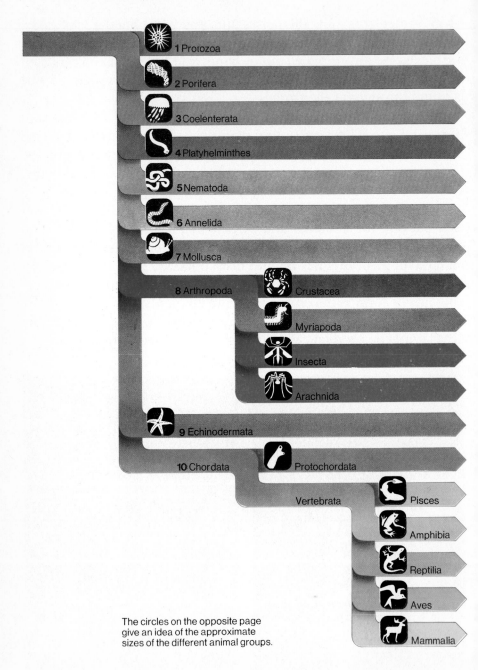

1 Protozoa
2 Porifera
3 Coelenterata
4 Platyhelminthes
5 Nematoda
6 Annelida
7 Mollusca
8 Arthropoda
  Crustacea
  Myriapoda
  Insecta
  Arachnida
9 Echinodermata
10 Chordata
  Protochordata
  Vertebrata
    Pisces
    Amphibia
    Reptilia
    Aves
    Mammalia

The circles on the opposite page
give an idea of the approximate
sizes of the different animal groups.

The protozoans are the first true animals. A microscope is needed to see most of them, but some are visible with a magnifying glass. Their bodies consist of a single cell and they live in moist surroundings such as seas, rivers, soil and inside other animals. They usually have a highly specialized way of obtaining food, but reproduce by a simple splitting (fission) into two or more new forms.

One of the best-known protozoans is the amoeba, the largest of which is no more than 0·5 mm across. It consists of a mass of jelly-like protoplasm surrounded by a delicate membrane, and constantly changes its shape. This is because it can squeeze out parts of itself into blunt projections called pseudopodia and it moves by means of these creeping fingers of protoplasm. In order to eat, it wraps itself around a tiny organism, digests what it can and leaves the rest. It reproduces every few days by splitting in half, each of which grows into a new amoeba. In unfavourable conditions, such as drought or cold, the amoeba can secrete a tough resistant coat around itself. The spore so formed can be carried by the wind. Most species of the amoeba live in ponds but some live inside the bodies of animals and

Greatly enlarged photograph of foraminiferans, a type of protozoan

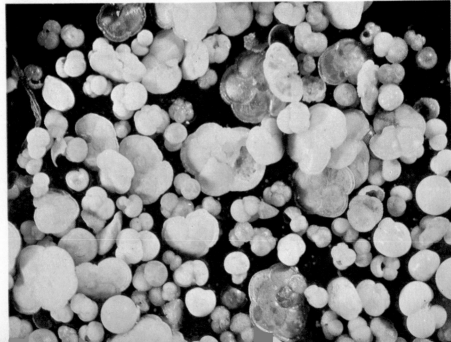

humans. Many of them are harmless but some cause disease, the most dangerous being a large type that causes amoebic dysentery.

Some amoeba-like protozoans manufacture skeletons or shells around themselves. The foraminiferans make their skeletons of calcium carbonate (chalk) and the radiolarians use silica, which is also the material of sand and glass. The pseudopodia of these animals protrude through tiny holes in the skeletons so that they can move about and capture food. Both these protozoans are found mainly in the sea, forming part of the drifting mass of small animals and plants called plankton, an important source of food for other sea animals. When they die their bodies decay and the hard skeletons sink to the bottom of the sea. Radiolarian skeletons form a muddy deposit called ooze, which covers the floor of the Pacific and Indian Oceans, an area of about 8 000 000 sq km. Foraminiferan remains also form ooze, but over the years some types are hardened and eventually become different kinds of rock. The white cliffs of Dover are made of chalk from countless foraminiferan shells, and some of the great Egyptian pyramids were built of limestone containing large foraminiferans.

Left: amoeba, one of the best-known protozoans. Right: mixed radiolarian skeletons made of silica

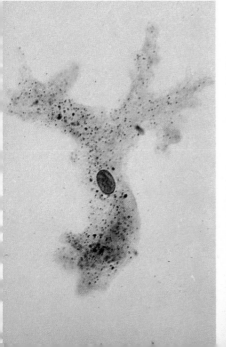 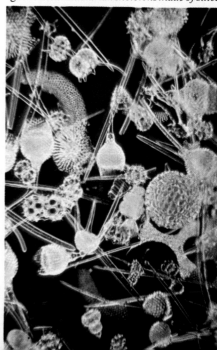

The ciliates are a large class of protozoans that have their outer surface covered with short hair-like projections (cilia). These move in a rhythmical wave, propelling the animals through the water and helping them to catch food.

One of the most common ciliates is *Paramecium*, which is found in many ponds. It is shaped like a rice grain with a rounded front and a pointed rear end and is about 0·25 mm long. The cilia propel it through the water so that it revolves like a corkscrew as it moves forwards. If it encounters any obstacles it moves backwards, changes its direction and moves on so that it avoids the obstruction. Unlike the amoeba, it has a fixed shape, and is covered with a rigid outer coat. It is also more complex, and has a 'mouth' consisting of a groove lined with cilia that waft particles of food into the animal. The food follows a definite path through the body and undigested waste leaves by a special opening. Like the amoeba, *Paramecium* reproduces by splitting into two parts but it also has a type of sexual reproduction, called conjugation. Two individuals first stick together and exchange hereditary material, then they separate and divide several times.

Left: *diagram of the parts of a* Paramecium. Right: *photomicrograph of* Euglena

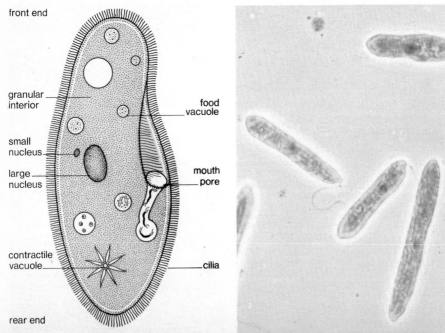

front end

granular interior

small nucleus

large nucleus

contractile vacuole

food vacuole

mouth pore

cilia

rear end

Stentor is a large trumpet-shaped ciliate with a ring of long cilia around its wider end which rotate and draw in food particles. It usually lives attached by its narrow end to a rock or other hard surface.

Another class of protozoans, the flagellates, are characterized by one or two whip-like structures called flagella, which are similar to cilia but much longer. The lashing of the flagella helps these organisms to move. *Euglena* is a flagellate that lives in ponds and resembles a plant in that it possesses the green pigment chlorophyll. This enables it to build up some of its food by photosynthesis, like plants. Near its front end is a red eye-spot which is sensitive to light. Some flagellates are parasites and cause diseases in animals and plants. One of the most harmful is *Trypanosoma* which causes African sleeping sickness and is spread by blood-sucking tsetse flies.

The Sporozoa, the fourth class of protozoans, contains parasites with a very simple structure. They reproduce by dividing to form many individuals, and can cause diseases in man and animals. One of the best known is *Plasmodium*, which causes malaria and is spread by the female mosquitoes of the genus *Anopheles*.

Photomicrograph of Trypanosoma *in human blood. It causes the disease sleeping sickness*

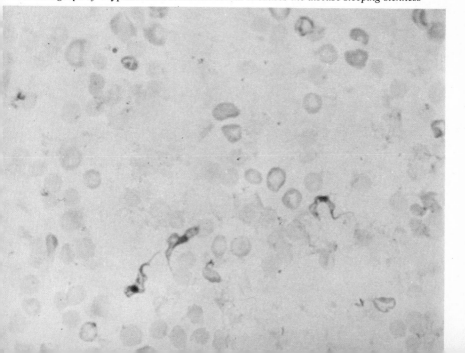

All animals except the protozoans are multicellular. That is, their bodies are made up of different types of cells which are specialized for different jobs. The phylum Porifera (meaning pore-bearers), which contains the sponges, forms one of the most primitive groups of multicellular animals and is probably not related to any other group.

Although most sponges live in the sea a few live in fresh water. All sponges are attached to hard surfaces such as rocks or the shells of other animals. Sponges have a hollow body with a large opening (pore) at the top and many smaller pores in the sides. The body is lined inside with special cells called collar cells, each of which has a collar-like funnel and a flagellum. The beating of all the flagella draws water in through the side pores (incurrent openings) and out through the top (excurrent opening). Food and oxygen in the water are extracted and absorbed by the collar cells. The body support of the sponge is a glassy, chalky or horny skeleton, which is the work of certain cells.

Sponges reproduce by buds, which separate from the parent and grow into new animals, and by branches, which grow horizontally

Left: *construction of part of a sponge colony.* Right: *same type of sponge as drawn in the diagram*

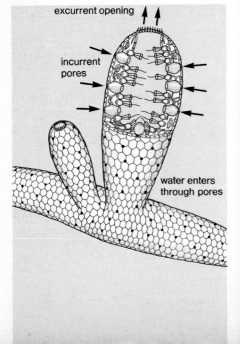

and give rise to new sponge colonies. Many sponges also produce buds (gemmules) which contain sponge cells and are covered by a taut outer coat. The gemmules can survive drought and freezing and release their cells under favourable conditions. Sponges can also reproduce sexually. Certain cells of the sponge develop into egg cells and sperm cells. A single egg fuses with a sperm cell and grows into a swimming larva. The larva eventually becomes attached to a hard surface and evolves to become a new sponge.

The types of sponge used for cleaning have a horny skeleton and are found in the warm shallow waters of the Mediterranean, the Gulf of Mexico and the Caribbean. Live sponges are caught, then beaten to loosen the protoplasm and left in the sun to dry and decay. The skeleton is then washed, bleached and trimmed. The ancient Greeks used sponges for washing and as padding for their armour. Sponges vary greatly in shape, size and colour. They may be rounded or branched, and range between one mm and one metre across. Sponges grow very quickly. If a small piece of living sponge is 'planted' in a suitable medium it will soon grow into a full-sized sponge colony.

Yellow sponge. The species includes many varieties, shapes and colours

Coelenterates are, like sponges, multicellular animals. They have a hollow body composed of two layers of cells which are separated by a layer of jelly. There is an opening to take in food and expel undigested remains. In many coelenterates this 'mouth' is surrounded by a ring of tentacles with stinging cells, which help to catch and convey prey into the mouth. Coelenterates are the lowest animals to have nerve cells. These are in the body wall, and impulses passing between them act as messengers, enabling the different kinds of cells to work together.

There are two types of body in coelenterates: the sac-like polyp, as in sea anemones and *Hydra*, and the flatter bell-shaped medusa, as in the jellyfish in which the jelly between the two cell layers is very thick. *Hydra* is a simple polyp living on stones or plants in fresh water. It is not easy to find because it is brown or green and although it grows to a length of three cm, it shrinks to a tiny blob if disturbed. It moves by attaching its tentacles to a firm surface and either looping or

Top: *extended polyps of a fan coral.*
Below: *jellyfish, showing bell and tentacles*

somersaulting along. The Portuguese man o' war is a large coelenterate consisting of a colony of polyps and medusas. Its sting is very painful and sometimes even fatal to swimmers. One type of medusa is specialized as a blue-tinted float and some polyps have tentacles up to 20 metres long. Jellyfish are found in nearly all seas and oceans. They swim by expanding and contracting their bell-shaped body. The largest recorded jellyfish was four metres across with tentacles 30 metres long, but the more common types such as *Aurelia* are usually between eight and 30 cm across.

Sea anemones are much larger polyps than *Hydra* and have many more tentacles around the mouth. They live attached to rocks in coastal regions, and a common anemone, *Actinia*, is found in rock pools. Corals are large colonies of polyps. Most types grow a hard skeleton made of limestone or a horny material. The best known are the corals of warm and tropical seas whose limestone skeletons form coral reefs such as the Great Barrier Reef of Australia.

Top: *sea anemones.* Below: *brown hydra, body and tentacles extended*

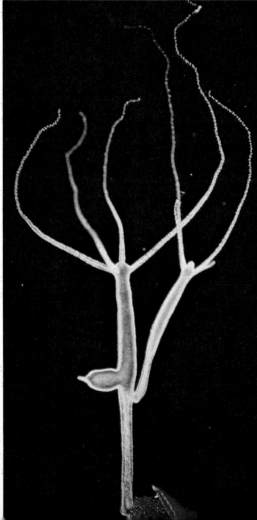

All animals except protozoans, sponges and coelenterates have bodies composed of three layers of cells. The phylum Platyhelminthes contains the flatworms, the most primitive of the three-layered animals. Flatworms have no blood system so that food and oxygen reach all cells of the body by simple diffusion. To make this easier the body is flat, and when there is a digestive system it has many branches. The phylum contains both free-living and parasitic forms.

Planarians are common free-living flatworms usually found in ponds, seas and moist soil. They have a brownish body with two simple eyes at the head end and a mouth in the middle of the undersurface. A tubular sucking organ (pharynx) protrudes from the mouth and food is obtained by sucking in the soft parts of other animals or dead animal matter. Most planarians have a complex reproductive system that produces both eggs and sperm. Many planarians can also reproduce by regeneration, during which the animal tugs itself into two, and the halves then grow the organs which each is lacking.

The flukes are a large group of parasitic flatworms, whose hosts

Left: *the head of a tapeworm has hooks and suckers for attachment to the intestines of the host.*
Right: *section of a liver fluke showing part of its excretory and reproductive systems*

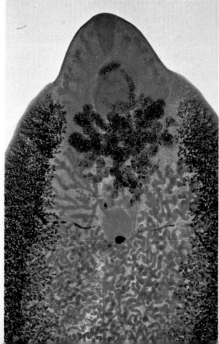

include man and many other vertebrates. They resemble planarians but have a sucker to cling to the host's tissues and a thick cuticle for protection. Blood flukes are found in human blood vessels and are particularly common in places such as China and Japan where people work barefoot in damp fields. When they enter the blood they undergo sexual reproduction and the fertilized eggs eventually pass out with the urine into the soil. The larvae then enter a water snail, undergo various changes, and pass out into the water where they may bore into a human foot and work their way into the blood system, so completing the cycle. One type, *Schistosoma*, causes the disease schistosomiasis in man. The liver fluke, *Fasciola*, causes severe damage and even death to sheep and cattle.

Tapeworms are long ribbon-like flatworms that are parasites in many vertebrate hosts. The body consists of a head with hooks and suckers, joined to a series of segments containing eggs in various stages of development. The tapeworm in humans passes part of its life cycle in the pig, and humans may become infected if uncooked pork is eaten which contains the tapeworm larvae.

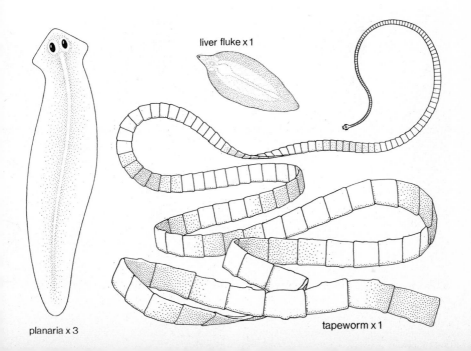

liver fluke x 1

planaria x 3

tapeworm x 1

The phylum Nematoda contains the roundworms. They are minute threadlike worms usually between 0·5 and 1·5 mm long, although some can grow as long as 30 cm. Apart from insects and protozoans they are probably the most widespread and numerous members of the animal kingdom, occurring in water and in most plants and animals. They are particularly numerous in the soil, where as many as 20 000 000 per square metre have been found. They feed on algae, bacteria, plant roots and other animals. Most are free-living in soil and water, but the best-known nematodes are the parasites of man, domestic animals, and plants such as potatoes and sugar-beet.

Roundworms are common parasites of man. One of the most common is *Ascaris*, which lives in the intestine and produces millions of eggs which pass out with the faeces. Where there are poor sanitary conditions, water or food contaminated with the eggs is swallowed and the eggs hatch into young worms in the intestine. A much more dangerous nematode is the hookworm, found where sanitation is poor and people go barefoot. The young worm burrows into the soft skin at the side of the foot and enters the blood. The adult worm,

Photomicrograph shows Bryozoa tentacles protruding from their protective cases to trap food

about 10 mm long, moves from the blood system and fixes itself to the inner wall of the intestine, where it feeds on blood and tissues causing loss of energy and decreased mental ability. Another harmful nematode is the filaria (genus *Wuchereria*). The adults grow up to 10 cm long and live in the lymph ducts of man, eventually blocking them. This results in the disease elephantiasis with enormous swellings of the lymph ducts. The Guinea worm (genus *Dracunculus*) is another harmful nematode.

The phylum Rotifera contains minute animals the size of protozoans but much more complex. At the head end is a crown of hairs (cilia), which beat like a revolving wheel and draw food particles into the mouth. Rotifers can often be seen in pond water when it is examined under a microscope. The phylum Bryozoa contains animals that look like seaweeds or mosses and are sometimes called moss animals. They live attached to rocks in fresh or salt water, often in large colonies, and they secrete an outer protective crusty skeleton. Despite a primitive appearance, they have characteristics which place them with the more advanced invertebrates.

The internal organs of a male rotifer show up clearly in a photomicrograph

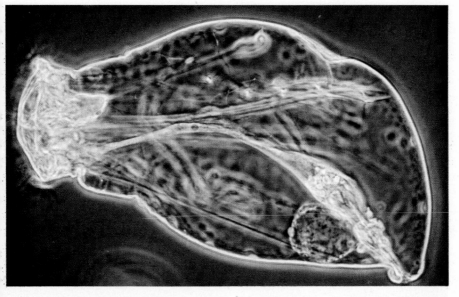

Molluscs are soft-bodied animals with a fold of skin (mantle) surrounding the body. They all have a muscular structure (foot) which in
gastropods is a flat sticky pad supporting the rest of the body. A space
between the mantle and the foot (mantle cavity) acts as a lung or, in
aquatic molluscs, contains gills which absorb oxygen. Many molluscs
have a protective outer shell secreted by the mantle. Some are highly
specialized for a particular way of life but most molluscs have the
same basic structure, which is different from any other animal and
has probably contributed to the spread and increase of the group.

   The gastropods, which are also called univalves, form the largest
group of molluscs and include the snails, limpets, whelks, peri-
winkles and slugs. They live both in water and on land and move by
means of the muscular foot. Snails have a distinct head with eyes on a
long pair of tentacles, which act like periscopes. Their digestive tract
and other organs are coiled within a spiral shell. Slugs are similar to
snails but have no shell. Snails feed on plant material by scraping off
fragments with a tongue (radula), which is covered with rows of sharp
teeth. Some gastropods are carnivorous. The whelk preys on

The Roman, or edible, snail is considered a great delicacy when cooked

barnacles and oysters and one type of slug eats other slugs. The brightly-coloured frilly sea slugs are mainly carnivorous.

Bivalves, such as oysters, scallops, clams and mussels, have two shells hinged together that open when the animals feed and move. Most bivalves live in the sea, and attach themselves to rocks like mussels, burrow in the sand like cockles and razorshells, or bore into rock like piddocks. Bivalves have large gills which trap particles of food and absorb oxygen from water which enters and leaves through a pair of openings called siphons. The food is trapped in mucus and carried to the mouth.

When bivalves are disturbed the two shells snap tightly together and are very difficult to prize open. The innermost layer of the shell is made of pearly calcium carbonate. Natural pearls are formed when this substance is secreted around irritating grains of sand inside the shell. Natural pearls are very rare, but pearls can be grown artificially by inserting tiny glass beads into the mantles of oysters. These are known as cultured pearls.

Blue mussels (left) *and scallops* (top right) *are good to eat.* Below: *sea slugs live in the depths of warm or tropical seas*

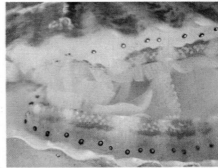

Cephalopods are the most advanced molluscs, with a well-developed brain and a pair of large eyes similar to those of higher animals. They all have the ability to learn. The octopus is particularly adept and can be trained to recognize different shapes and find its way through a maze. Most cephalopods have no outer protective shell although some have an inner shell. The body is covered by a muscular sac (mantle), which contains the gills and other organs. Cephalopods can move by jet propulsion, taking water into the cavity between the mantle and the body and expelling it forcefully through a tube (funnel) projecting from the body. The funnel can be pointed either backwards or forwards and so controls the direction of movement. Cephalopods catch food with their tentacles and tear it up with a pair of beak-like horny jaws. Most of them can expel a cloud of inky black liquid which acts like a smokescreen to confuse their enemies. They can also change the colour of their skin very rapidly to produce camouflaging colours and patterns.

The ammonites are an extinct group of cephalopods which were common in Jurassic times, about 150 000 000 years ago. They occur as fossils in rocks today. The only surviving relative of the ammonites is the pearly nautilus, which has a spirally-coiled shell and many sticky tentacles. Squids and cuttlefish have a ring around the mouth of eight short tentacles bearing suckers, and two longer club-shaped ones to catch the fish and crustaceans on which they feed. A porous bone-like substance embedded in the mantle acts as a buoyancy chamber and supports the body. This is the familiar cuttlebone given to pet birds. Most squids are only a few centimetres long, but the giant squids in the northern waters of the Atlantic Ocean are the largest living invertebrates. Many have a body length twice the height of a man and a total length, including tentacles, of up to 15 metres.

Octopuses are less active than squids and have only eight tentacles. They lie in wait for their prey and then dart out to catch it. Most of them are quite small but some grow up to 10 metres across. The female lays her eggs in capsules, which are deposited in cavities in rocks. The eggs, tended by the mother, eventually hatch into larvae, which form part of the plankton before developing into adults.

mantle

funnel

octopus

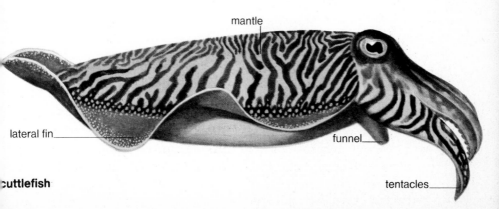

mantle

lateral fin

funnel

cuttlefish

tentacles

The phylum Annelida contains earthworms, bristle worms and leeches. The body of an annelid is made up of a series of rings called segments. All annelids have a fluid-filled cavity between the outer body wall and the digestive canal, so that the outer body wall can move independently of the rest of the body.

Earthworms vary in size: the common earthworm, *Lumbricus terrestris*, can grow up to 20 cm long and five mm wide, while the giant Australian earthworm can be as big as three metres long and two cm wide. *Lumbricus* has about 150 segments and its lower (ventral) surface is flatter and paler than its upper (dorsal) one. The skin, through which it breathes, is kept supple and moist by slimy secretions of mucus. There is no obvious head, but there is a mouth at the front end of the body, on the ventral surface, and an opening (anus) at the rear end. A pale swollen section called the saddle (clitellum) occurs about 30 segments from the front. Most segments have four pairs of bristles (chaetae) on the sides and underneath, which grip the ground when the worm is

Earthworms are the most common worms

moving along. Although it has no proper eyes it can distinguish between light and dark and it also has a simple brain. Earthworms have a system of blood vessels through which red blood is pumped by five ring-like hearts.

Earthworms are hermaphrodite, that is, they have both male and female reproductive organs, but they cannot fertilize their own eggs. The clitellum produces a sticky fluid which holds the two worms together during mating, when the mutual exchange of sperm takes place.

Earthworms feed mainly on dead plant material and often emerge on damp nights and pull leaves into their burrows. When burrowing some earthworms, but not *Lumbricus*, swallow large quantities of soil, digest the plant material in it and eject the rest, which appears on the surface as wormcasts. By this action earthworms help to keep the soil in good condition. They break up the soil, mix in plant material, bring mineral-rich soil to the surface, and increase the oxygen content. There may be over a million worms per hectare (10 000 square metres) of land.

A cast of coiled earth produced by an earthworm

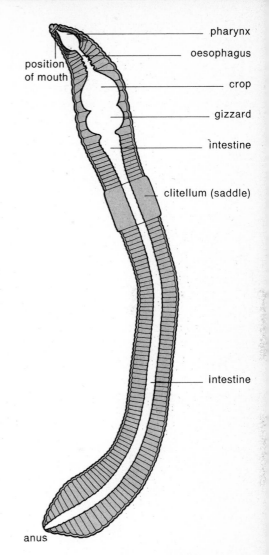

digestive system of earthworm

Although bristle worms look like earthworms, most of them live on the sea-shore, either in burrows or in tubes built of sand, stones or other materials. They have a row of limb-like lobes down each side of the body, each ending in a bunch of bristles (chaetae). The lobes are used for swimming. The ragworm, *Nereis*, about 20 cm long, is found mostly in coastal waters. It feeds on smaller animals caught with a pair of pincer-like jaws. Some bristle worms have unusual methods of sexual reproduction. The palolo worm, which lives among the coral reefs of the Pacific, grows special segments at its rear end containing eggs and sperm. At a certain time of the year this section breaks free and swims to the surface where it sheds its contents, and fertilization takes place between the eggs and sperm of different worms. The remaining part of the worm grows more reproductive segments the following year.

Another common coastal worm is the lugworm, *Arenicola*, which has much smaller swimming lobes than the ragworm and spends all its time burrowing in the sand. It feeds by swallowing

Top: *leech*. Below: *fanworm or featherworm*

sand, digesting the organic material in it, and ejecting the rest. Its casts are often seen at low tide on sandy beaches. Many worms live in tubes which they build of sand or mud and from which they can extend their tentacles to catch food and absorb oxygen. These tube-dwelling worms include the fanworm and the strange-looking peacock worm whose long brightly-coloured tentacles form a crown surrounding its head. It retreats into its tube when it is disturbed.

Most leeches live in fresh water or damp marshy land. The body of a leech, with suckers at both ends, has fewer segments than that of other annelids and does not have chaetae. Most leeches are quite small but a few reach a length of 30 cm. They move by swimming or looping along using their suckers. Leeches feed by sucking the blood or other body fluids of larger animals, including man. A leech can rapidly take in as much as three or four times its own weight in blood and can live on it for weeks. The freshwater leech, *Hirudo*, was used by doctors until well into the 20th century to bleed patients.

The lugworm is a sand-burrower

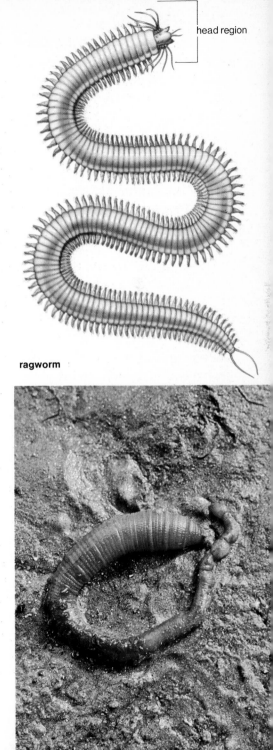

head region

ragworm

The arthropods form the largest phylum in the whole animal kingdom, numbering about eighty per cent of all the animal species. The phylum includes four classes: the crustaceans which include lobsters, crabs, barnacles and woodlice; the arachnids which include spiders, scorpions and mites; the myriapods which include centipedes and millipedes, and the insects, which are the largest group.

The body of an arthropod is basically a series of segments, each of which bears a pair of jointed limbs (appendages). In the higher arthropods, however, many of these segments have become fused together or specialized for a particular function. The body is usually divided into head, thorax and abdomen, and in some arthropods the head and thorax are fused into a cephalothorax. In primitive arthropods the appendages are all alike, but in more advanced forms they are modified to form swimming paddles, fighting weapons and food crushers. Their feelers (antennae) are sensitive to touch, heat and chemicals (smell). Arthropods have a firm tough outer covering (exoskeleton) which is made chiefly of a horny substance called chitin. The exoskeleton protects and supports the soft parts of the animal.

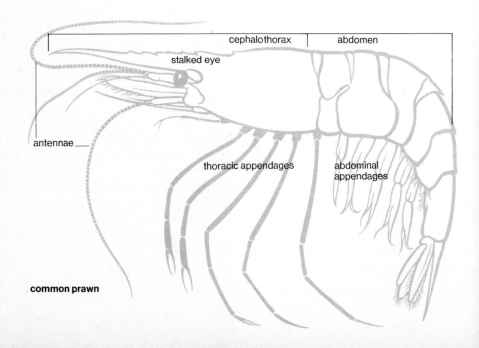

common prawn

Unlike the bones of vertebrates, which are made of living cells and can grow, the exoskeleton of an arthropod is secreted by cells in the body wall and cannot grow. Therefore it is shed periodically to allow the animal to grow in a spurt before the new skeleton has time to harden. At that stage arthropods are in great danger from their enemies because their soft bodies are unprotected. Although their muscular systems differ, arthropods move like vertebrates because their walking legs, like the limbs of vertebrates, are jointed, and the exoskeleton provides a firm attachment for the muscles that move the legs.

Arthropods have an elaborate body enclosed within the exoskeleton. They have an efficient nervous system, and a blood system in which the blood is contained not in tube-like vessels, but in large cavities so that it bathes the organs. Most arthropods have well-developed eyes, especially the insects, which have compound eyes made of hundreds of separate sections. There is usually one eye on each side of the head, often on a movable stalk, but there may also be other simple eyes on top of the head.

Left: *compare the hinged joint of a human limb* (top) *with that of an arthropod limb* (below).
Right: *the giant centipede of Malaya makes side-to-side progress using its jointed limbs*

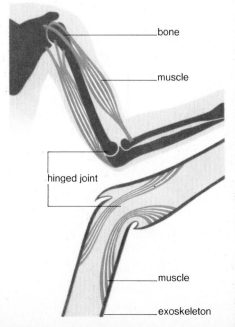

bone

muscle

hinged joint

muscle

exoskeleton

Almost all crustaceans live in water; either in the sea, like lobsters, crabs and shrimps, or in ponds and streams like freshwater crayfish and waterfleas. Some tropical hermit crabs and the giant robber crab live on land several kilometres from the coast, but they must still return to the sea to breed. All crustaceans breathe by means of gills.

The common lobster, *Homarus vulgaris*, is a good example of a large crustacean, although it is not as big as some of the giant crabs that may measure over a metre from claw to claw. Lobsters are found around the coasts of Europe and North America. Their shell (exoskeleton) is blue-black in colour, with a protective shield (carapace) covering the head and thorax. They have 19 pairs of limbs. These include a large pair of claws for catching prey and fighting enemies. There are also four pairs of walking legs and a number of small appendages used for swimming. At the end of the lobster's body is a fan-shaped tail. Usually lobsters walk or creep over the sea-bed, but they can also swim backwards very quickly by jerking the tail. During reproduction the eggs are fertilized as they emerge from the body of the female by sperm from the male. The fertilized eggs remain

The common lobster, which lives in cold seas, is a favourite meal for many people

attached to the female and eventually hatch into free-swimming larvae. Lobsters and crabs are fierce fighters and greedy feeders and they will devour carrion as well as living prey. If one of a lobster's limbs is held by an enemy or caught between rocks, it will pull away and leave the limb behind; eventually another one will grow. Hermit crabs live inside the empty shells of whelks and similar molluscs. They have no exoskeleton on the hind part of their soft bodies and therefore seek protection in the shells of other animals. Closely related to the hermit crab is the robber crab, *Birgus*. It is said to climb coconut palm trees but probably eats only those coconuts that have fallen to the ground. It uses its pincers to chip away the wood of the coconut and extract the meat. Crustaceans are good at concealing or disguising themselves both from their enemies and their prey. The shells of hermit crabs are often camouflaged with seaweeds and sea anemones.

Woodlice, which look rather like very small armadillos, are the best known of the few land crustaceans. Their shells are not waterproof, and to avoid the danger of drying up they must live in very moist surroundings. Their gills have been modified to breathe air.

Top left: *common crab*. Below left: *woodlouse*. Right: *hermit crab examines periwinkle shell*

In addition to the big crustaceans such as crabs and lobsters there are thousands of smaller types, many of which are too small to be seen without the aid of a microscope. One of the most common and most interesting is *Daphnia*, a waterflea about three mm long and therefore fairly easy to observe. *Daphnia* is found in fresh water and is not a true flea (fleas are insects), but it swims with a sort of jumping movement that resembles the hopping of a flea. Its body is enclosed in a transparent shell, so that when a live *Daphnia* is observed under a magnifying lens, all its organs are visible, including the digestive tube and pulsating heart. It has one large eye and five pairs of legs which are used to direct a continuous stream of water into its mouth, where food particles are filtered out and eaten. It swims by means of its large antennae. Another small and very common freshwater crustacean is *Cyclops*, which has a single eye at the centre of its head. It belongs to the copepods, most of which are found in the sea.

Waterfleas and similar small animals are very important in freshwater life because they form a major source of food for fishes and

Biology students often study Daphnia *because it is a common crustacean and its structure can be easily seen under a microscope*

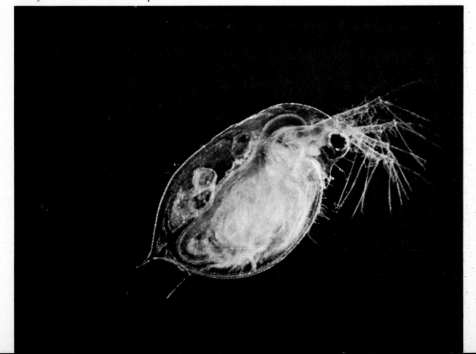

other larger animals. They themselves feed on microscopic plants and therefore form an essential link in many food chains.

Barnacles are, in fact, crustaceans although they look like molluscs. They have a thick chalky shell like many molluscs but also have the jointed legs typical of crustaceans. Their larvae (young stages) are also characteristic of crustaceans. Most barnacles are hermaphrodites (each animal has both male and female sexual organs). The free-swimming larva undergoes changes and eventually becomes attached to a solid object, where it remains throughout its adult life. Acorn barnacles live attached, by means of their head end, to the bottoms of ships and rocks on the sea-shore. They 'kick' food into their mouths by means of six pairs of curled legs. Barnacles are so numerous on the hulls of ships that they must be scraped off to prevent fouling. Goose barnacles have a stalk made partly from the head, and are found on rocky shores. Barnacles are so well protected and equipped for the life they lead that they are found by the million along any stretch of rocky coast.

Goose barnacles (left) *live in clusters on sandy shores. Acorn barnacles* (right) *are a nuisance to sailors because they attach themselves to the hulls of ships*

Arachnids are terrestrial arthropods in which the body is divided into two main sections, the cephalothorax, or head section, and the abdomen. The cephalothorax bears special jaws called chelicerae, a pair of leg-like feelers and four pairs of walking legs. Unlike insects and crustaceans, they have simple eyes only and no antennae. Their main method of breathing is by organs called lung books. These are outgrowths of the body wall, looking like the pages of a book, that occur in cavities on the underside of the body and absorb oxygen.

Spiders are generally disliked by man, but although the bite of some spiders is poisonous and can be very painful, it is often no more dangerous than a bee sting. The bite of the tarantulas, including the large South American bird-eating spider which can be 18 cm across, is not usually serious to man. The most poisonous of the spiders is the female black widow spider of the Americas, but even the bite of this species is not always fatal. The female spiders tend to be bigger and more poisonous than the males; the male in some species is killed by the female after mating.

Spiders usually eat insects that are regarded as pests by man. Some

Left: *garden spider catching and eating a fly*. Right: *bird-eating spider*

spiders hunt their prey, but most trap them in webs spun with a sticky silk thread produced by organs (spinnerets) at the tip of the abdomen. The webs are delicate and complex and different patterns are spun by different types of spider. As soon as the insect is caught in the web, the spider hurries out and kills its prey by injecting poison and digestive juices into it and then sucks up the dissolved body of its victim. If the prey is large the spider may first bind it up in silk.

Many spiders also use the silk for other purposes. Newly-hatched spiders climb up high from their birthplace and spin a thread which acts as a parachute to carry them to a new habitat. The water spider spends all its time in ponds, living within a kind of diving bell made of silk which contains air brought from the surface. The trap door spider builds a burrow covered with a hinged top made of earth and silk. It then darts out and grabs any passing prey. The female spider of most species lays her eggs in a silk cocoon which she either hangs on the web or carries about until the young hatch. Some spiders even carry their young around for a short time.

Top left: *water spider*. Below left: *a berry-like East African spider*. Right: *the hunting spider weaves a cocoon for her young*

Scorpions are ferocious-looking arachnids. Some of the tropical species grow up to 25 cm long but most are much smaller. They are not usually aggressive to man but they sometimes attack other scorpions. They have powerful pincers near the head and a hooked sting at the end of the flexible tail which can be whipped upwards and forwards over the back. They hunt at night, seizing prey with their pincers and paralyzing it with the sting. The female produces living young, rather than eggs, and she carries them around on her back for a week or more after birth.

The smallest and most common arachnid is the mite, which has an oval body between one mm and two cm long. Soil mites are very important in helping to keep the soil fertile by breaking down dead plant material, and there may be hundreds of millions of them per hectare. Others eat fungi or small animals. Some mites live in water while others are parasites such as the itch mite, which causes mange in cattle and scabies in man. The red spider mite is a well-known plant parasite, causing great damage in apple, plum and pear orchards.

Top left: *scorpions carry their young on their backs.* Below left: *a scorpion eats a grasshopper.* Right: *red spider mites attach themselves to a harvestman*

Another group of mites feeds on dead organic matter, and is a pest of flour, cheese and other human foods. Ticks are similar to mites but a little larger and feed on the blood of man and of animals such as cattle and sheep. Many of them carry and transmit fatal diseases which cause great damage among domestic animals.

Another common type of arachnid is the harvestman, which looks like a spider with very long thin legs and a small round undivided body. Unlike spiders it does not produce silk.

A very strange primitive arachnid is the king crab or horseshoe crab, *Limulus*, which lives near the Atlantic coast of North America and in Asia. It is not a true crab but a survivor of an ancient group of arachnids which has existed for hundreds of millions of years. This living fossil has a heavy dark shell, which can reach the size of a dinner plate, and a jointed tail 20 cm long. The spidery legs and other parts are completely hidden under the shell. It preys on worms and bivalves on the sea-bed at night and remains hidden with only its eyes above the sand by day. It breathes by means of gill books, which are very similar to the lung books of spiders and scorpions.

king crab

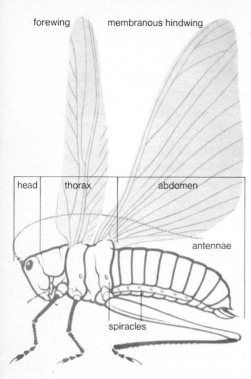

forewing membranous hindwing

head | thorax | abdomen

antennae

spiracles

There are over 850 000 named species of insects in the world, which means they form about three-quarters of all animal species. All insects are relatively small; the rhinoceros beetle, one of the largest insects, is only 15 cm long. The largest fossil insect ever found was a dragonfly with a wingspan of nearly a metre. All insects have a body divided into three parts: the head, thorax and abdomen. The head bears antennae, three pairs of jaws modified for the particular feeding method, and a pair of compound eyes. The thorax has three pairs of legs and up to two pairs of wings.

Most insects live on land. They breathe by means of branching air tubes (tracheae) which run through the body and carry oxygen to all parts. Each trachea has an opening (spiracle) on the side of the body. Movements of the body help to push air backwards and forwards in the tracheae, but such a system of breathing is only suitable for small animals, which is one reason why insects have never evolved into large organisms. Those insect larvae (young stages) which live under water

Top: *diagram shows the structure of an insect.*
Below: *silverfish are familiar household pests*

breathe by gills.

Zoologists divide insects broadly into those with and those without wings; there are many more species of winged insects. Insects differ in their type of metamorphosis which is the series of changes that occur during development from the young form (larva) to the adult (imago). Some species have complete metamorphosis: the larva is very different from the adult and undergoes complex changes, including the formation of an intermediate stage (pupa), before becoming an adult. Other insects have incomplete metamorphosis, in which the larva resembles the adult.

The wingless insects are the most ancient and primitive types and are usually very small. The silverfish is about 1·5 cm long. The springtails, Collembola, are very common and live in the soil, where they eat dead plant material. They are only one-third to five mm long, and move by means of a forked 'tail' which is curved under the body and can project the insect forward about 30 cm – the equivalent of a man doing a long-jump of about 300 metres.

Top: *the horsefly has a compound eye.*
Below: *North American luna moth*

The Lepidoptera is a large order of insects containing the butterflies and moths, which all undergo a complex metamorphosis during development. The adults have two pairs of large wings covered with overlapping scales. The mouth parts are modified to form a long tube (proboscis) which is used for sucking nectar from flowers and sap from fruit and can be coiled like a watch spring when not in use. The larva of butterflies and moths is the well-known caterpillar, which is quite different in appearance from the adult. It has powerful jaws and feeds on plants, often causing great damage to crops. In some caterpillars the spiracles can be seen as a series of dots along the sides of the body.

There are more species of moths than butterflies. Butterflies usually settle with their wings closed over their backs. They have antennae with a club-shaped knob at the tip, and they fly by day. Moths, on the other hand, fly mostly at night or twilight, have a thicker body and less colourful wings which are held in various positions while they rest and they have antennae of various shapes.

Butterflies and moths of temperate countries usually have a wing-

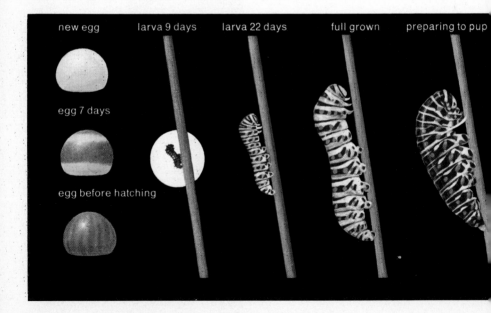

new egg   larva 9 days   larva 22 days   full grown   preparing to pup

egg 7 days

egg before hatching

span of between two and six cm, but tropical species, such as the bird-winged swallowtail, have a span of up to 20 cm. In many species, the wing markings and colour make an excellent camouflage for protection against other animals. The males are often more brilliantly coloured than the females and many species are great favourites of collectors.

Most butterflies and moths fly fairly slowly but many can fly great distances. Some species migrate. The monarch butterfly of North America, for example, flies south 2000 km or more to warmer regions in vast swarms each autumn. It rests over the winter in trees, then flies north again in the spring.

Most butterflies and moths are harmless, but the caterpillars of some are serious pests to growing and stored crops. One useful moth is the silkworm moth, *Bombyx mori*, of China. Its caterpillar feeds on mulberry leaves and spins a cocoon of silk fibres for the chrysalis (pupa). The silk can be collected and used commercially. Attempts to use the giant North American silkworm have failed because the threads are not continuous and the cocoons cannot be unravelled.

pupating   chrysalis   imago emerging   fully emerged

The largest order of insects is the Coleoptera, containing about 275 000 species of beetles and weevils. All beetles have a shiny appearance, because the forewings are modified to form a pair of smooth tough wingcases (elytra) which protect the delicate hindwings, used in flying, and the rest of the body. Beetles can fly, but are more usually found on the ground in soil, under bark or among plants. They have strong jaws and eat almost any type of food. Beetles vary greatly in size. The smallest may be less than $\frac{1}{2}$ mm long, but the largest, the rhinoceros and hercules beetles, grow to a length of 15 cm.

The well-known glow-worms and fireflies are beetles. The females have luminous organs on their undersurface, giving out a soft greenish light which attracts the males. Even stranger is the rare South American beetle, *Phrixothrix*, whose larva is called the railroad worm because it has a red light on its head and a row of greenish ones on its sides.

Some beetles are harmful because they feed on crops and stored food. The common Euro-

Top: *scarab beetle.* Centre: *Colorado potato beetle.* Below: *red and black beetle*

pean cockchafer causes great damage by stripping acres of trees bare of their leaves, and its larvae kill crops by eating the roots. The Colorado beetle, common in the United States, destroys the leaves of the potato and related plants. Weevils, which all have the head drawn into a long snout, feed on cotton, peas, apples and many other crops. The deathwatch beetle and furniture beetle bore into and feed on wood, causing great damage.

Not all beetles are harmful. Ladybirds eat aphids and scale insects which are regarded as pests by man. Ground beetles such as the small bombardier beetle have very long legs to hunt down their prey, and often feed on soil pests. Burying beetles bury the dead bodies of animals much larger than themselves by scraping away soil from beneath them. The females then lay their eggs in the dead animal, which is used as food by the young larvae when they hatch. In a similar way the dung beetles lay their eggs in dung. The scarab, which is related to the dung beetle, was worshipped by the ancient Egyptians.

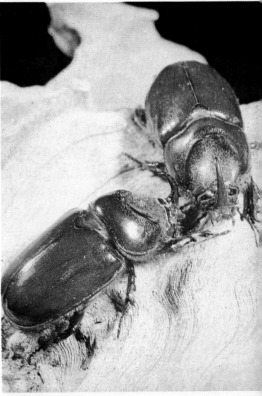

Top: *female dung beetle*. Centre: *male rhinoceros beetles*. Below: *glow-worm*

The large group of insects which contains the wasps, ants, bees, saw-flies and ichneumon flies is the Hymenoptera. They all have two pairs of delicate transparent wings and mouth parts adapted for biting, licking or sucking up nectar. The females have egg-laying tubes (ovipositors) which may be adapted as stings, drills or saws.

The ichneumon fly is a parasite and lays its eggs in caterpillars. By killing a large proportion of the caterpillars hatched each year it plays an important part in controlling leaf-eating pests. Some of its relatives are parasites of plants and may cause galls (oak apples) on oak trees. The galls contain the developing grubs. Wood wasps differ from the typical wasps in having no 'waist' between the thorax and the abdomen. The ovipositor of wood wasps is adapted for boring through wood, in which the female lays her eggs.

Many hymenopterans are social insects – they live in groups, with each member specializing in certain jobs. The common wasp, *Vespa*, is a social insect and lives in large colonies in nests built of a papery substance made of chewed wood. It does not store food, and therefore the whole colony, apart from the young hibernating queen (an egg-

The female wood wasp has a specially adapted ovipositor for boring through wood

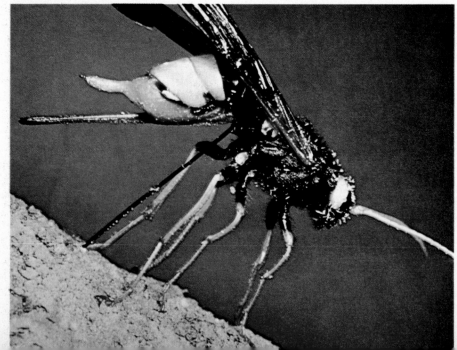

laying female), dies in the autumn. In the following spring the queen starts producing eggs which will develop into the new colony. She must feed the first batch of grubs herself, but as more wasps develop they go out and collect food, while the queen spends all her time laying eggs. In the early part of the year wasps feed on insects and grubs; later, however, they feed mostly on fruit.

There are about 6000 species of ants in different parts of the world. An ant colony is divided into three castes: the queen, the males and the wingless workers which are sterile, non-egglaying females. The males and queen fly off from the nest to mate, forming the swarms of flying ants sometimes seen in summer. After mating the males die and the queen loses her wings and sets up a new colony.

Some tropical ants, especially the army ants of tropical America, travel from place to place in huge columns, eating anything in their way. Other ants make use of different insect species to benefit the colony. Some ants 'keep' aphids and feed on the sweet liquid (honey-dew) which these insects excrete. The robber or slave-making ant forces ants of other species (the slaves) to do all the work of the colony.

Many species of aphids excrete a sweet liquid on which ants like to feed

There are many different types of bees but only a few are social insects that live in colonies. The best-known social bee is the honey bee, *Apis mellifera*, which has been kept by man since ancient times. Social bees live in a colony or hive which contains one egg-laying female (queen), about 50 000 females (workers) that are sterile and cannot lay eggs and a few hundred stingless males (drones). Drones develop from unfertilized eggs. All the females develop from fertilized eggs; some larvae which are fed only on royal jelly, a substance secreted by the workers, develop into queens. Most of the larvae are fed mainly on pollen and nectar and grow into workers. The queen spends all her time laying eggs, a thousand or more a day throughout the summer, and she places each egg in a waxy comb cell. The drones have only one job: they fertilize the queen on her mating flight and then die or are killed by the workers. All the eggs laid by the queen in her three or four years of life are the result of this single mating.

Worker bees spend their first two or three weeks of life looking after and feeding the larvae, then they begin to produce wax for building more cells and also act as hive guards. Only the second half

The bee grubs develop in the cells of the hive

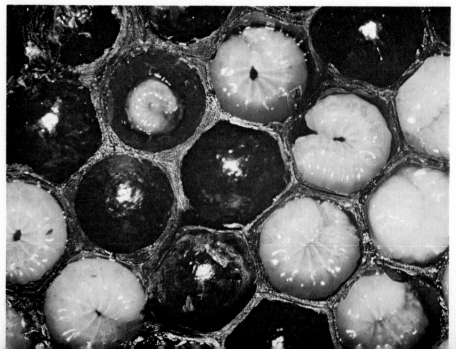

of their six-week life is spent collecting pollen and nectar. The pollen is carried in special baskets on the hind legs, and the nectar is swallowed, then brought up again (regurgitated) on returning to the hive. The nectar not required immediately is converted into honey, which is stored for winter use. After collecting nectar and pollen, bees always return to their own hive rather than to any other. They are thought to navigate by recognizing landmarks and by using the sun as a compass. Bees also have a 'language' of their own. One bee can communicate with the rest of the hive by means of special types of movement known as dances, which are thought to indicate where a plentiful source of food can be found.

Throughout the winter the queen and workers remain in the hive, drowsy but still needing food. The bee-keeper must therefore feed them with sugar to replace the honey he has taken. In summer, if a hive has become overcrowded, the queen flies off with many of the workers to start a new colony. In the old hive there are new queens developing, and the first to emerge will immediately kill her sisters in their cocoons, then set out on her mating flight and return to lay eggs.

Left: *worker bees build wax storage cells for the honey.* Right: *a honey bee alights on a flower*

pollen basket

clover flower

There are about 85 000 species in the order Diptera (two-winged insects), which includes flies, mosquitoes and midges. In all of them the second pair of wings has become reduced to a pair of tiny balancing organs, important for maintaining steady flight. All dipterans feed by sucking up liquid. The housefly sucks up fluids from rotting material or dissolves sugary food with saliva before sucking it up. Mosquitoes, tsetse flies and horseflies are blood-suckers and are harmful because they carry disease-causing micro-organisms in their saliva. When saliva is injected to stop the blood they are sucking from clotting, the disease micro-organisms are transmitted to the host. Some tropical mosquitoes carry malaria and yellow fever, and tsetse flies carry sleeping sickness which also affects cattle. Flies can contaminate food from the rotting material on which they feed and breed. Many other blood-sucking dipterans carry disease or cause skin irritation, and the larvae of many flies are parasites of plants and animals.

Many dipterous larvae are leg-

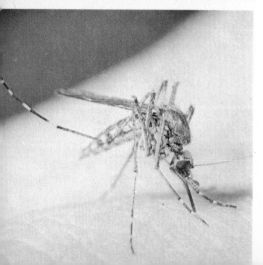

Culex mosquito. Top: *pupae and larvae.*
Centre: *male emerges from pupa.* Below: *adult*

less grubs (maggots) which live as parasites or saprophytes in rotting animals or plants. The larvae of mosquitoes and midges are very active and live in water. They feed there on tiny plants and breathe through a minute tube pushed up to the surface. Mosquitoes are controlled by oil sprays which kill the adults and suffocate the larvae.

Fleas are wingless insects but are otherwise similar to the two-winged flies. They are in a separate order, the Siphonaptera. It is thought that their ancestors had wings. They live by sucking blood from warm-blooded mammals and birds. Their body has flattened sides to squeeze between the hairs or feathers of their hosts. They are all powerful jumpers. The larvae live on animal matter in and around the host's nest or home which is why animals constantly on the move do not have fleas. Their main danger to man is the spreading of disease. The Black Death of the 14th century and the Great Plague of 1665 were the same disease, which was spread when rats carrying the fleas were transported by trading ships.

Top: *blowflies.* Centre: *housefly larvae and pupae.* Below: *large flea attacks African rat*

A large group of insects have incomplete metamorphosis: there is no pupa stage and the larvae (nymphs) resemble the adults in form and habits. Dragonflies belong to the order Odonata. The dragonfly nymph is aquatic and carnivorous, living in ponds and streams and breathing by gills. It takes as long as five years to complete its development. The adult, however, lives for only about a month. Like the nymph, it is carnivorous, and feeds mainly on flies. Mayflies (order Ephemeroptera), have three-pronged tails. Their aquatic nymphs live a long time, may moult as many as twenty-three times, and feed on plant life. The adult mayflies cannot eat or drink and live only from a few minutes up to a day.

The order Orthoptera includes locusts, grasshoppers and crickets. Locusts are serious pests in many warm countries as they can form enormous migrating swarms which consume all plant life in their path. A medium-sized swarm can contain as many as 1 000 000 locusts covering over 100 sq km. Because a locust eats its own weight of food a day, the swarm would devour about 3000 tons of food daily.

Left: *snouted harvester termites guard the nest as workers repair it.* Right: *fungus-growing termite mound, five metres high, found in Africa. Mound shapes vary with the different soils*

Each female deposits several pods in moist sand. Each pod contains about 100 eggs, and the wingless nymphs (hoppers) pass through five stages before they become flying adults. Migrating hoppers are killed by sprays or poisoned food and the flying adults are sprayed with insecticides from the air.

Termites, also called white ants, look like ants but belong to the order Isoptera. They live in large colonies in tropical and warm countries. The male (king) lives with the female (queen) for several years, periodically supplying her with sperm for the eggs, which she lays at a rate of up to 30 000 a day. Her abdomen becomes an enormous egg bag over 10 cm long, making her completely helpless. The other members of the colony are sterile. There are workers who tend the colony and feed the young, and soldiers who have large heads and jaws and attack unwelcome intruders. Some termites build huge nests of earth or sand which are twice as high as a man and contain about 500 000 insects. Termites are serious pests when they attack buildings or furniture because they can digest any plant material, including wood.

Left: *desert locust*. Right: *mating dragonflies*

The echinoderms are a group of marine animals that includes sea urchins, starfish, feather-stars and sea lilies, brittlestars and sea cucumbers. The body of an echinoderm is symmetrical and its parts are in fives or multiples of five, such as the five arms of a starfish and the five sections of a sea urchin. The soft parts are protected by a hard skeleton (test), which is formed inside the body. Also within the body is an important series of water-filled canals that connect with each other and function like a blood system to transport dissolved food and oxygen. Protruding from the surface of the body are many cylindrical extensions of the canal system (tube feet) which can be extended and retracted for movement and feeding.

Sea urchins are rounded or heart-shaped animals whose test forms a rigid capsule enclosing the body. The test is covered with many sharp spines which help the animal to move about. Urchins feed on vegetable matter and particles of dead animals. Starfish are less rigid than sea urchins, with five or ten movable arms radiating from a central disc. The test consists of many chalky plates embedded in the

Left: *mating starfishes.* Right: *oral (underside) view of a sea urchin, showing the suckered tube feet and the centrally-located mouth*

body wall, forming a firm but flexible covering. Starfish feed on clams, oysters and similar animals. A starfish will prise open the shell of its prey by using its tube feet and then turn its own stomach inside-out and envelop and digest the exposed body of its victim. Starfish and sea urchins live in rock pools and shallow seas.

Brittlestars are echinoderms with long thin arms joined to a tiny central disc. They tend to break very easily but can grow new parts successfully. Sea cucumbers are strange worm-like animals with a test of tiny spikes embedded in the body wall. The body wall itself is covered with a thick, rather slimy layer of skin, and the mouth is surrounded by feathery tube feet used in feeding. The crinoids are the most primitive group of echinoderms. They have feathery arms and the mouth is directed upwards from the body. Sea lilies are crinoids that were much more common millions of years ago than they are now; they are often found as fossils today. They live attached to a hard surface by means of a stem. The most common type of crinoid today is the feather-star which, unlike sea lilies, can move about.

The sea cucumber has a softer body than other members of its phylum. Front tentacles snare small animals, and the tube feet act as suckers for attachment to the sea bottom

The phylum Chordata contains all animals that at some stage of their life possess a notochord. This is a rod made of gristle that extends down the length of the animal and stiffens and supports the body. Chordates also have gill slits and a hollow nerve cord. The gill slits are used by fish in respiration and by the lower chordates in feeding, but are reduced and non-functioning in higher vertebrates.

The great majority of chordates are animals with backbones and are placed in a separate subphylum, the Vertebrata. In these animals the notochord is present only in the embryo. In the adult the notochord is enclosed within a much stronger supporting structure, the backbone, which is made of bone or cartilage. Vertebrates also have an entire skeleton made of bone or cartilage, which gives greater support and protection to the internal organs and much more efficient powers of movement. Vertebrates have a well-developed nervous system, with a brain protected by the skull and a spinal cord protected by the backbone.

Chordates that are not vertebrates are sometimes grouped together as protochordates. They are usually tiny marine animals and, despite their relationship to vertebrates, seem very primitive.

The sea squirt is a protochordate which lives in the sea and is sometimes found in rock pools on seaweed. The adult has a jelly-like body covered with a transparent outer coat with two openings through which water enters or leaves, food and oxygen being extracted on the way. There is no trace of a notochord and only a simple nervous system. Most sea squirts have no powers of movement so that when disturbed, the body contracts and squirts out jets of water. The larva (ascidian tadpole) of the sea squirt, on the other hand, is an active swimmer and has a long tadpole-like body with the notochord and nerve cord extending along its length.

The most advanced of the protochordates is the lancelet, *Amphioxus*, in which the notochord is present in the adult. The lancelet is about four cm long and is found in many parts of the world, spending most of its life burrowing in sand in shallow water.

The adult red sea squirt has two body openings through which water enters and leaves. It thrives in salt water. Although the larvae can swim, the adult becomes attached to rocks

The most primitive vertebrates are the lampreys and hagfish. They belong to a group of animals, the Cyclostomata, which resemble fishes except that they have no jaws; they are characterized by a row of gill pouches, used for respiration, down each side of the body just behind the head. Lampreys and hagfish are very similar, with an eel-like body covered by a layer of slime and a single fin near the tail. Their history goes back a very long time; fossils have been found which are thought to be the ancient representatives of the group that lived about 400 000 000 years ago.

Lampreys are found throughout the world and have long been used as food by man. The Romans kept them alive in tanks and served them as delicacies and old records tell us they were sold at fairs and markets in Britain. King Henry I of England was popularly thought to have died from a surfeit of them.

The sea lamprey of the Mediterranean and northern Atlantic waters grows to a length of one metre. Most of its life is spent in the sea, but it makes its way up-river to spawn. The brook lamprey, which is less than half as long, lives all its life in fresh water. Spawning takes place

Lampreys feed by attaching themselves to their prey and cause great damage to other fish

in the spring and the eggs are laid and partly buried in sand. The eggs hatch into worm-like larvae which are blind and toothless. They remain in this form for up to six years, and are so unlike the adults that for a long time biologists thought they were different animals.

Lampreys have no jaws, and the mouth is a large circular funnel-shaped sucker situated at the end of the head. This sucker and tongue are set with a number of sharp horny structures which resemble teeth but have a different structure from the true teeth of fishes. Lampreys feed by attaching themselves to their prey, usually fish, by means of the sucker and then suck the blood and rasp the flesh off their victim by moving their tongue backwards and forwards. They cause great destruction in shoals of fishes and fishermen regard them as dangerous pests.

Hagfish are similar to lampreys. They have a worm-like body, about 90 cm long, which is covered with slime produced by special glands along the sides of the body. A hagfish will rasp its way into the body of the fish on which it is feeding, consume the muscle and other internal parts, but leave the skin intact.

Left: *close-up showing a lamprey's suckers, hooks and teeth.* Right: *lamprey attached to a trout*

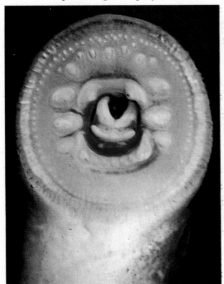

Fishes are vertebrates adapted for a life spent entirely in the water. They have a smooth streamlined body which is protected by a water-proof skin and usually covered with scales. The scales are sharp and toothlike in sharks and related fishes; in bony fishes they are thin and horny. Fishes feel slimy to the touch because special glands in the skin are constantly secreting mucus onto the surface of the body. The scales help to prevent loss of water from the body in salt water and dilution of body fluids in fresh water. In fishes such as salmon, the scales are circular and overlap like the tiles on a roof. They have growth rings, one for each year of life, which enable us to tell their age.

Most fishes obtain the oxygen they need for breathing from the water by means of gills. These are delicate tissues, rich in blood vessels, which are situated on bony or gristly supports in slits along both sides of the throat. While the fish gulps a mouthful of water and forces it through these slits, oxygen is extracted from the water and absorbed into the blood vessels of the gills. At the same time carbon dioxide passes from the blood and out of the gills to the water.

All fishes have fins, which enable them to move through the water

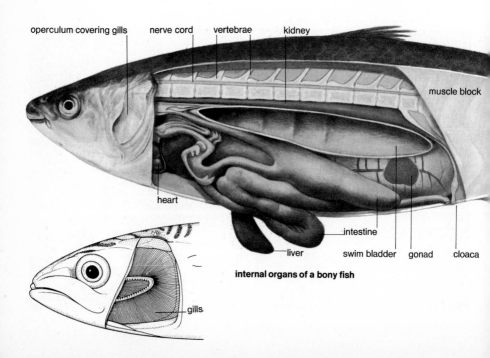

internal organs of a bony fish

and keep a steady balance. Two pairs of fins, called pectorals and pelvics, which correspond to the limbs of land vertebrates are used for steering, balancing and forward movement. In addition, there are unpaired fins along the back, the belly and at the end of the tail, which provide the main forward thrust in locomotion.

Both cartilaginous and bony fishes originated in the Devonian period about 390 000 000 years ago. Both groups have evolved to become very successful and widely distributed in their way of life. Many cartilaginous fishes have become specialized as fast-moving predators. The bony fishes have undergone even greater adaptation so that they are now specialized for living in all sorts of conditions in the water. Most inhabit the warm sunlit layer near the surface, others live in the middle depths, and a few are adapted for living in the deepest part of the ocean where little light or warmth penetrates.

Some fishes can live for a time out of water; the climbing perch and mudskipper, for example, use their pectoral fins to move about on land and the flying fishes use their pectorals like wings to skim over the surface of the sea.

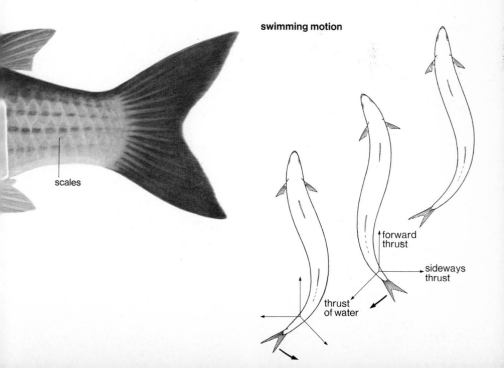

swimming motion

scales

forward thrust

sideways thrust

thrust of water

Biologists classify fishes into two broad groups: cartilaginous or gristly fishes (class Chondrichthyes) and bony fishes (class Osteichthyes). Cartilaginous fishes have a skeleton made of cartilage. They can be easily recognized because they have visible gill slits behind the eye, the gills do not have a gill cover and the tail has one lobe larger than the other. They do not have an air bladder inside the body to keep them buoyant, so when not actually swimming they tend to sink. Their skin is very rough to touch, being covered with sharp backward-pointing spines or scales (denticles) which are very similar in structure to their teeth.

The main types of cartilaginous fishes are sharks, skates and rays. Sharks are the great flesh-eaters of the ocean, feeding on fish of all sizes as well as animal plankton. They have always been greatly feared by man and several species will attack swimmers. If there are traces of blood in the water they will furiously attack anything within range, even wounded sharks. Their sense of smell is extremely well developed and they can follow a scent as a bloodhound does on land. Whale

Left: *a mermaid's purse with an embryo shark inside its egg case.* Top right: *shark swimming* Below right: *a rare action photograph of a manta ray*

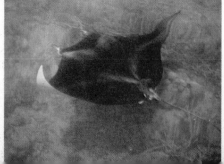

sharks and giant basking sharks are the biggest fish in the world, growing to a length of 15 metres and weighing more than 12 tons. They are harmless to man and feed only on plankton. The dogfish is a small shark known as rock salmon. Like a dog, it has a keen sense of smell and hunts in packs appearing in schools near the shore. Sharks produce large eggs which are fertilized inside the body of the mother and, in many species, develop in a horny container called a mermaid's purse. In some sharks the purse stays inside the mother so that the young shark is born alive rather than hatching from the purse. In others, the purse is deposited in the sea by the mother and the young shark eventually hatches from it.

Rays have flat bodies with long tails and swim by flapping their wing-like pectoral fins. They may grow to a large size, the manta ray weighing over a ton. In spite of their fearsome appearance only a few species are dangerous. The sting ray can inflict a painful wound, and the electric ray stuns its prey and enemies with an electric shock of up to 80 volts. Skates are very similar to rays, but usually have smaller pectoral fins and their tails lack the spines often seen in rays.

Some sharks attack humans but one of the largest, the sand shark, is harmless to man

The bony fishes (class Osteichthyes) are the most familiar to us. They have a skeleton made of bone, not cartilage as in cartilaginous fishes (class Chondrichthyes). Bony fishes which are good for eating include the herring, cod, salmon and trout, but the class also contains sea horses, eels, goldfish and most aquarium tropical fishes. Most bony fishes are members of a large group, the ray-finned fishes, with fins made of webs of skin supported by many slender horny rays. They are the most widespread group of fishes living in the sea, fresh water and estuaries, from the arctic to the equator, and from the deepest to the upper sea levels. Only a few members still exist of two other groups of bony fishes: the lobe-finned fishes (crossopterygians) and the lungfish.

Bony fishes can be easily distinguished from cartilaginous fishes. In all except primitive types, such as the sturgeon, the gills are protected by a covering flap so that the individual gill slits are hidden, and the tail is two equal lobes. The body is covered with overlapping scales or, in some groups, bony plates, arranged in such a way that the skin does not feel rough when stroked from head to tail. Almost all bony fishes produce great numbers of eggs; for instance, one female sturgeon can lay 3 000 000 eggs during her life. The eggs are poured out of the mother's body and fertilized in the water by sperm from a male fish. This is called external fertilization and is the most usual method of reproduction in aquatic animals. Cartilaginous fishes, however, reproduce by internal fertilization (the eggs are fertilized inside the mother's body). We eat the eggs and sperm of bony fishes as roe: the eggs form the hard roe and the sperm the soft roe. Caviar is the processed hard roe of the sturgeon.

Unlike cartilaginous fishes, bony fishes have a swim, or air, bladder. This air-filled chamber just above the gut acts as a buoyancy tank. It is carefully adjusted in various ways so that the fish can float at a particular depth. If a deep-sea fish is suddenly brought to the surface the air in its bladder expands and causes the bladder to burst. It is thought that the swim bladder evolved from a primitive kind of lung, which also developed into the lungs of lungfish and amphibians.

Kingfish are a widespread tropical species of bony fish – this school was found in the Indian Ocean

The migration of fishes is one of the most fascinating aspects of nature. What drives them to make such long journeys, and how they find their way back to the exact spot where they were born, is still a mystery.

Not all fishes migrate, but among those that do are the salmon. They spawn in the upper reaches of rivers in North America and northern Europe. After two or three years in fresh water the young fishes swim down to the sea, where after about a further four years they grow into mature adults. Then they turn homewards, sometimes swimming 500 miles to spawn in the same streams in which their parents spawned years earlier. This is the most dangerous time of their lives, battling against the river currents and leaping up rapids. Natural enemies prey upon them, and fishermen wait to catch them. Finally, a fraction of the thousands that set out actually reach the end of their journey, thin and worn out. The female uses her tail to sweep a shallow trough in the river-bed in which to lay her eggs, and the cycle is completed.

Zoologists took many years to piece together the story of the eel,

stages in the life of a salmon

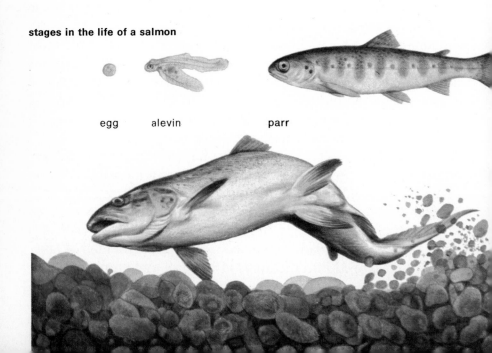

egg   alevin   parr

another fish that migrates. The European eel lives in rivers and coastal waters throughout western Europe and the Mediterranean. It feeds mainly at night, devouring almost anything – worms, fish spawn, crustaceans, frogs, dead voles and water birds. Every autumn great numbers wriggle and writhe their way through mud and wet grass, from stream to river, and down to the sea. Once there, they set out on a 3000- or 4000-mile journey across the Atlantic Ocean to their spawning ground south-east of Bermuda in the deep waters of the Sargasso Sea. The parents die when they have laid their eggs. The young hatch out in the spring. They have a thin leaf-like body and do not resemble adult eels at all. These minute creatures begin their long trip to European waters, drifting and swimming, aided by the Gulf Stream. The journey takes over two years, and during this time they gradually become elvers, a more eel-shaped form, and make their way up-river. The related American eel breeds in the same area, but swims west instead of east. The sturgeon also migrates. Spawning takes place in rivers where the young spend the next two years of life before migrating to the sea.

smolt

Many fishes are interesting because they have specialized body forms, colour, habits and way of life. In several species parents take great care of their young. Some fishes, for example, carry their young in their mouths, dropping them only to feed, then picking them up again. The male three-spined stickleback builds a nest, courts a female with a special dance and chases her into the nest to lay the eggs. He will vigorously defend his territory from other sticklebacks and when the eggs hatch he brings food for the young and looks after the nest. Once the female stickleback has laid her eggs she takes no further interest in the eggs or young. The male Siamese fighting fish is brilliantly coloured with large beautiful fins which spread out during courtship of the female. This little fish is also extremely fierce and will fight ferociously with another male of its species. However, he is also a very good parent and guards the young until they are old enough to take care of themselves.

Perhaps the fiercest fish of all is the South American piranha. It is about 25 cm long and resembles a carp with large razor-sharp teeth. Piranhas swim in large shoals and usually feed on other fishes, but if a

An angler fish prepares to catch its prey

wounded animal or man falls into the water the blood attracts the shoal, and they can very quickly strip all the flesh from the victim.

The most colourful fishes are found in tropical waters. In the bright sunlit water and among the coral reefs their vivid colours and patterns act as camouflage to protect them from enemies or make them unnoticeable as they hunt for their prey. The angel fish, a typical example, has thin dark lines on its fins and body which look like the plant stems around which it lives. The angel fish continually waves its fins and body, imitating the movements of the plant stems. The frog fish and sea dragon swim among seaweed and have fins, spines and fleshy processes which look like seaweed and disguise and protect them from their enemies. The plaice and other flatfish live at the sea bottom, flat against the sea-bed. They have spots or blotches which break up the outline of their bodies and make them inconspicuous. The angler fish, living in the depths of the sea, has a long spiny process which curves forward from the top of its head, acting like a line and bait. Small fishes will approach, attracted by the bait, and are snatched and swallowed.

Top left: *piranha*. Below left: *Siamese fighting fish*. Right: *angel fish*

Amphibians evolved from fishes about 350 000 000 years ago. But how could animals which have adapted to living partly on land have developed from those so perfectly adapted to life in water? A few clues are found by studying some of the fishes living today.

The Indian climbing perch travels from pond to pond overland, using its spiny gill covers, fins and tail to 'walk'. It breathes air, and has a pair of lung-like organs as well as gills. The mudskipper, found in tropical countries, especially the coastal regions of Africa, moves with the aid of fleshy pectoral fins which function as limbs on land. It can also use its limbs to climb up plant stems and mangrove roots in search of food.

The lungfishes are a small group of freshwater fishes that live only in Africa, South America and Australia. They have lungs richly supplied with blood vessels which enable them to breathe air. The African lungfish, *Protopterus*, lives in swamps and streams and can survive seasonal droughts by burying itself in mud at the bottom of a stream and surrounding itself with a cocoon of mucus, which covers the whole body except the mouth and prevents the fish from drying

A mudskipper uses its fleshy pectoral fins to climb out of the water onto a twig

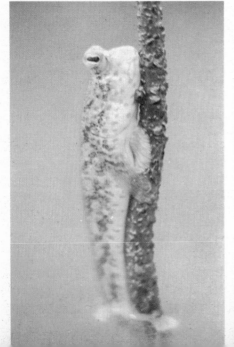

out. It makes a small air-hole which leads from its burrow to the surface of the mud so that it can breathe.

It is thought that amphibians evolved from primitive bony fishes called crossopterygians, which first became practically extinct about 70 000 000 years ago. Most of them lived in fresh water. They all had fins in the form of fleshy lobes like the mudskippers, and some of them also had lungs. Zoologists think that they used their fins in a way similar to the mudskippers to crawl from pond to pond during droughts, because they could breathe air with their lungs. The earliest amphibians were very similar to crossopterygians, and probably evolved from them. As the amphibians evolved still further, the lobed fins of the crossopterygians gradually developed into the true walking limbs seen in amphibians today. Until recently it was thought that crossopterygians were extinct. But in 1938 a living marine representative of this group, called a coelacanth, was discovered off the coast of South Africa. It was a large fish, about 1·5 metres long – a good example of a 'living fossil'. Since then, several more coelacanths have been discovered.

The coelacanth is the only surviving crossopterygian

The class Amphibia is a small group of vertebrates that includes frogs, toads, newts, salamanders and caecilians. The group was much larger in later Palaeozoic times when there were many amphibians, most of which looked like a somewhat smaller version of the modern crocodile. Amphibians were the first group of vertebrates to adapt to life on land, and it is thought that they evolved from lobe-finned (crossopterygian) fishes. Their structure is midway between fishes and reptiles. Like reptiles they have adapted to terrestrial life by developing two pairs of limbs, lungs for breathing air and nostrils opening into the mouth cavity. Amphibians differ from reptiles, however, by having a moist skin which is used, in addition to the lungs, in respiration. All adult amphibians must return to the water to breed, and most of them also need to live in moist surroundings even when they are not breeding. The eggs and larvae (tadpoles) of amphibians develop in water, and the tadpoles have gills and other adaptations to an aquatic life. They undergo metamorphosis when developing from the larval to the adult stage. This means that the body changes from

Left: *spotted salamander.* Right: *the male oak toad is shown here singing at night to attract females for mating*

being adapted to an aquatic life to being adapted to a life spent partly on land. Amphibians are carnivorous, feeding on small insects and other animals. The skin is poisonous in some species while in others it is specially developed for carrying the eggs around until the tadpoles develop.

Of the modern amphibians, the urodeles (newts and salamanders) are probably most like the ancestral forms, but are much smaller. They have tails and long slender bodies which in many tropical species are brightly coloured. Newts tend to be smaller than salamanders and, even as adults, spend much time in the water. Salamanders have stouter bodies, usually brightly coloured, and most of them spend much of their adult life on land. The largest amphibian is the giant salamander of China and Japan, which is almost two metres long. Some salamanders, such as the Mexican axolotl, retain the features of their larval stages and spend the whole of their life in the water. The strangest amphibians are the caecilians – burrowing animals found in the tropics that have no legs and look like large earthworms.

Left: *the palmate newt is smaller than the salamander.* Right: *an albino axolotl displays its feathery external gills*

Frogs and toads form a group of amphibians (anurans) without a tail. The body is short and squat, and the hind legs, which are long and very strong, have become adapted for hopping on land. Frogs and toads range in length from about one cm to about 24 cm. The largest are the African goliath frog and the South American giant toad. Giant toads and certain South American frogs have a poisonous skin. Frogs and toads have large eyes which are very sensitive to fast-moving prey. They shoot out their long sticky tongues, which are attached to the front of the mouth, very quickly to catch the flies, worms and other small animals on which they feed. Their teeth are used for holding the food rather than chewing. Tree frogs have sticky pads on their toes to help them climb and jump along the branches of the trees in which they live. Like many other frogs the tree frog has a very loud and unmusical croak. The North American bullfrog has a booming bull-like croak and one type of large South American frog mews like a cat.

A frog will usually breathe by taking air in through its nostrils by the rapid raising and lowering of the floor of its mouth. The mouth

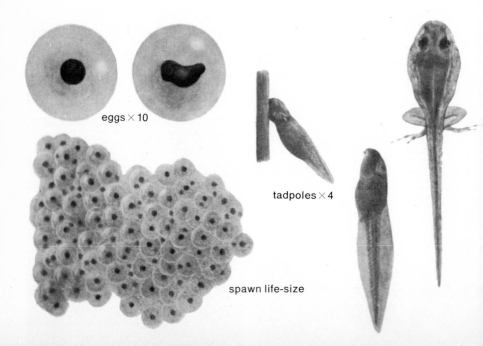

eggs × 10

tadpoles × 4

spawn life-size

cavity is richly supplied with blood vessels through which oxygen is absorbed and carbon dioxide is given out. Sometimes the frog will give an extra-large gulp, forcing the air into its lungs where a similar exchange of gases occurs. Frogs and toads also exchange gases on the surface of their moist skin.

Frogs breed in ponds at springtime. The smaller male climbs onto the back of the female and external fertilization takes place; the male deposits the sperm over the eggs as they are laid. Each egg is surrounded by a layer of protective jelly which swells on contact with water to form the well-known masses of frogspawn. The spawn of toads is similar but produced in long strings. Each egg hatches into a tiny tadpole which soon develops external frilly gills, eyes, mouth and tail and feeds on waterweed. Later the external gills are replaced by fish-like internal gills and the tadpole then feeds on small animals. Eventually legs and lungs develop and finally the tail shrivels and disappears; it is re-absorbed into the body and provides food for the final metamorphosis of the tadpole into a tiny frog or toad. The young adult finally comes to the surface of the water and crawls onto land.

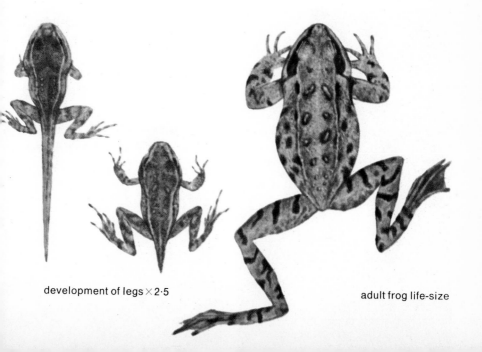

development of legs × 2·5

adult frog life-size

Reptiles and amphibians look very alike, and in fact early zoologists classified them in the same group. But there are important differences between them. The skin of most reptiles is dry and covered with scales, while that of amphibians is moist and unscaly. Terrestrial reptiles live the whole of their lives on land and most of the aquatic species come ashore to breed, but amphibians always return to the water to breed. This means that the reptile egg has an outer shell which is tough and leathery to prevent the embryo from drying out. When the egg hatches, the young reptile resembles the adult; there is no larval stage as in amphibians. Reptiles and amphibians are both cold-blooded – they cannot control their own body temperature which goes up and down according to the temperature of the surroundings. They differ from mammals and birds in this way.

In Mesozoic times the reptiles formed a very large group which included the mighty dinosaurs, but today the class includes only tortoises and turtles, lizards, snakes and crocodiles. Tortoises, turtles and terrapins belong to the order Chelonia, which contains about 250 species. The bodies of the chelonians are enclosed in a protective

Left: *red-eared terrapin*. Right: *Indian star tortoise with yellow pattern on its shell segments*

shell through which the head, legs and tail protrude.

Because of their structure, tortoises can move about only very slowly. However, they are so protected by their shell that they do not need to move quickly to escape predators. The shell is in two parts, connected by bridges between the legs, and consists of a layer of bony plates covered by a layer of horny plates. It is the horny layer of the hawksbill turtle of tropical and subtropical seas that is used to make tortoiseshell combs and jewellery. The shell of the Indian star tortoise has a yellow star-like pattern on each of its segments.

The limbs of land chelonians end in claws, whereas those of aquatic species are specially developed as paddles for swimming. In Britain, marine chelonians are called turtles, land types are called tortoises and freshwater forms are called terrapins. In America, however, all chelonians are usually called turtles.

Turtles and tortoises can grow to an enormous size. The giant tortoises of the Galapagos Islands and the leatherback turtles are the largest living chelonians, sometimes reaching a length of two metres and a weight of 450 kg.

A marine turtle from the Ascension Islands uses its limbs as paddles

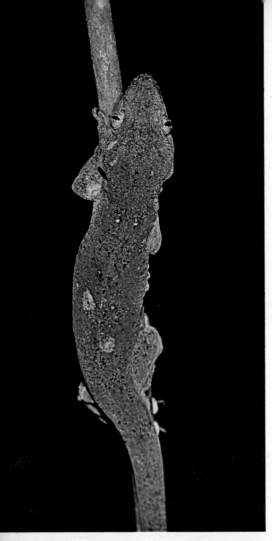

Lizards are the largest group of living reptiles. They are very common in warm and tropical regions but become rarer near the poles. Lizards are found as far north as Lapland and as far south as the tip of South America. They have a long, soft, scaly body and a long tail, and most of them have four legs except a few which are legless. Lizards shed their tails when seized by an enemy; the rest of the creature escapes and a new tail is soon grown.

Iguanas form a large family of lizards that are mainly found in North and South America. They are often large and brightly coloured and have a crest of horny spines along the back. The group includes the marine iguanas of the Galapagos Islands which were of special interest to the English naturalist Charles Darwin. The agamid lizards are another large family, very similar to the iguanas, which live in Africa, Asia and Australia. The flying dragon of south-east Asia has wing-like flaps of skin down each side of its body to permit a gliding movement from tree to tree. The geckos are small, noisy,

Top: *rain forest gecko.*
Below: *female slowworm with her young*

mostly nocturnal lizards found in all warm regions. They often run about on the smooth walls and ceilings of houses by means of special pads on their toes. Most chameleons live in Africa. They are well adapted for living in trees, with tail and toes specialized for grasping, and a skin that rapidly changes colour to blend with their surroundings. The eyes of a chameleon can swivel around independently of each other in search of insect prey, which is then caught by the long sticky tongue. Slowworms are burrowing lizards without legs. They are covered with smooth shiny scales and live under logs and stones. The small European lizards all belong to the same family (Lacertidae). The most common British lizards are the sand lizard, the green lizard and the wall lizard.

The largest living lizard, found only on a few islands in Indonesia, is the Komodo dragon, a monitor lizard which grows up to three metres long. The only poisonous lizard is the brightly-coloured Gila monster which inhabits United States desert regions.

Top: *agamid lizard*. Centre: *chameleon*.
Below: *common iguana*

Snakes are actually legless lizards which are highly specialized for burrowing, climbing or swimming. A very flexible backbone allows the snake to make a side-to-side movement, and the thick scales on its belly help it to grip rough ground and push itself forwards. These scales also help a snake to move in a straight line in the same way as a caterpillar. At one time it was thought, quite incorrectly, that a snake walked on its ribs. Many snakes swallow their prey alive but some have quick and efficient means of killing it first. Constrictors, such as boas and pythons, squeeze their victims to death by wrapping their strong bodies around them. The pythons of south-east Asia, India, Africa and Australia are very large and non-venomous, and have vestiges of hind legs, a primitive feature in snakes. Venomous types have special hollow poison fangs through which venom is injected into the prey. The forked tongue of a snake is not poisonous: while flicking in and out it acts as a sensitive feeler. Vipers are venomous snakes of Europe, Asia and Africa. Unlike

Top: *puff adder*. Centre: *grass snake*.
Below: *Eastern diamondback rattlesnake*

most other snakes, their young are not hatched from eggs but born alive. The adder, a type of viper, is Britain's only venomous snake and its bite, although not usually fatal, can be quite serious. The puff adder, a very large African viper, inflates its body to double its size when disturbed. The venomous rattlesnakes, which live in North and Central America, have a rattle at the end of the tail made up of horny plates which are vibrated to produce a high-pitched buzz. This warns off enemies and saves the venom for killing prey.

Snakes can swallow prey much bigger than themselves because their very mobile jaws are connected by elastic ligaments. Snakes have no ear-drums and cannot hear airborne sounds. They can, however, sense vibrations such as footsteps. Cobras used by snake-charmers are probably hypnotized by the movements of the charmers' hands; they certainly do not hear the music. These snakes have a characteristic habit of expanding the hood on their neck when they are annoyed.

Top: *spectacled cobra with hood expanded, ready to strike.* Below: *D'Alberti's python*

Reptiles of the order Crocodilia (crocodilians) are the closest living relatives of the dinosaurs. Crocodilians, found in freshwater habitats in tropical and subtropical regions, are all very similar to each other and differ only in minor points of structure. Crocodiles are found in all continents except Europe. Alligators live in the swamps of the southern United States and in southern China. They have a broader snout than crocodiles, and their teeth do not overhang the jaw. Caimans occur in South and Central America and the very long-nosed gharials are found in India.

All crocodilians have four short limbs with webbed toes and horny claws, and a long muscular tail for swimming and to stun their prey. Their tough leathery skin is made up of rough oblong scales and, unlike lizards and snakes, they do not moult. Their strong and very long jaws are rimmed with conical teeth for gripping and crushing prey, rather than chewing it. Their nostrils and eyes are set high up on the head so they can lie almost completely submerged and still be able to see and breathe. The nostrils can be closed by special muscles when the animals dive. The ears of crocodilians are protected by scaly flaps

Left: *a young crocodile in the River Nile photographed at night.* Right: *crocodile farm, Malaysia*

which are raised to expose the ear-drums when above water and lowered when the animal is submerged.

Crocodilians often bask in the sun by day and hunt in the evening and at night. They can swim swiftly but are clumsy on land. They generally eat fish but will also feed on large land animals, which they pull into the water and hold under until they drown. The largest crocodiles are quite capable of killing a man. Many grow to six metres or more, and may live for longer than 50 years.

After mating, the females lay between 20 and 90 eggs, each measuring up to 10 cm long, and covered with a hard shell. The eggs are laid in a hole in the sand or in a nest of rotting leaves and branches. The mother stays near the eggs until the young hatch, after about two months, having been incubated by the heat of the sun or rotting vegetation. The young crocodilians break the shell with the aid of a temporary egg-tooth on the end of the snout. They are about 20 cm long at birth, but grow rapidly. In some parts of the world crocodiles and alligators have become very scarce because their skins are hunted and there are special crocodile farms where these reptiles are bred.

Alligators live in swamps of the southern United States and southern China

Reptiles were the dominant animals during most of the Mesozoic era, from about two hundred and twenty five million to sixty five million years ago. The best known are the land-living dinosaurs, divided into four main groups. The largest group contains the sauropods – four-legged dinosaurs that fed on plants and had very long tails and long necks. The 26-metre long *Diplodocus* was the longest sauropod and the 50-ton *Apatosaurus*, more popularly known as *Brontosaurus*, was probably one of the heaviest. Some of these dinosaurs are thought to have been partly aquatic; their bodies were so clumsy they may have found it easier to move about in water. The second group consists of the armoured dinosaurs which had a tough scaly body covering and were smaller than the sauropods, but, like them, had four legs and fed on vegetation. The eight-metre long *Stegosaurus* had a double row of upright plates on its back and spikes on its tail. The slightly smaller *Triceratops* resembled a rhinoceros with three horns on its head and a hard bony armour covering both head and neck. Another group of herbivorous dinosaurs, the ornithopods, were bipedal – they used only their hind legs for walking. The

Pterodactylus 600mm wingspan

Diplodocus 26 m

Ichthyosaurus 3 m

best known is *Iguanodon*, over 10 metres long. The flesh-eating dinosaurs belonged to the theropods, all of which were bipedal. The largest, *Tyrannosaurus*, had a huge head with teeth up to 15 cm long.

There have been various theories to explain why dinosaurs became extinct. At the end of the Mesozoic era there was a change in climate which led to changes in vegetation and surroundings. The horsetails and evergreens, which formed the main diet of the plant-eating dinosaurs, were replaced by modern deciduous trees. In addition, the land was rising so that the seas in which the aquatic dinosaurs lived were drained. The dinosaurs, highly specialized for living in the climate and conditions of the Mesozoic, could not adapt to these changes which was probably one of the main reasons for their becoming extinct. Some Mesozoic reptiles were adapted for living in the sea. The plesiosaurs had long necks and grew to a length of 7·5 metres. The ichthyosaurs resembled dolphins and whales and reached a length of about three metres. There were also flying reptiles, the pterosaurs. The best known are the pterodactyls, which had bat-like wings and short hind legs. The largest, *Pteranodon*, had a wingspan of nine metres.

Tyrannosaurus 12 m

Stegosaurus
8 m

Plesiosaurus 7·5 m

Birds form a widespread group of animals well adapted to their way of life, which is based on flying. The other large group of flying animals are the insects. A bird's whole body is organized to provide quickly the great amount of energy needed for flying. All birds have a covering of feathers, which are not only essential for flight but also prevent body heat loss. They are also warm-blooded, which means that their body temperature and metabolism are steady and do not rise and fall with the temperature of their surroundings. The skeleton of a bird is very light and strong and the bones contain many air cavities. There is a large ridge down the length of the breastbone, to which the powerful muscles needed for flying are attached. The clavicles – bones that form the collar bones in man – are fused into a V-shape: this is the well-known wishbone of a chicken. Tarsals and metatarsals, the bones that make up the ankle and foot in man, are fused in birds into a single bone, the tarso-metatarsus. In a similar way, the carpals and metacarpals, forming the wrist and hand in man, are reduced to carpo-metacarpals in birds.

Birds have a keen sense of sight and hearing; a protective tuft of feathers covers their ears and in many species cannot be detected. The bill (beak) is a very characteristic feature; it always has two nostrils on the upper surface, but the shape varies with different species. The digestive system is characterized by a thin-walled crop, in which food is stored before being passed to the thick-walled gizzard. Birds have no teeth, so food must be ground up in the gizzard before passing to the rest of the digestive system.

The digestive and reproductive systems open into one chamber (cloaca), with a single external opening (vent). Birds excrete solid uric acid rather than urine; this prevents too great a loss of water from the body. Fertilization is internal, the sperm passing from the cloaca of the male to that of the female. Like reptiles, birds lay eggs and in most species one or both parents incubate the eggs by sitting on them. Another reptilian feature is the scaly covering of the legs. Birds vary in size from the ostrich, which is about two metres tall, and the condor, with a wingspan of nearly four metres, to the hummingbirds, some of which are only five cm long.

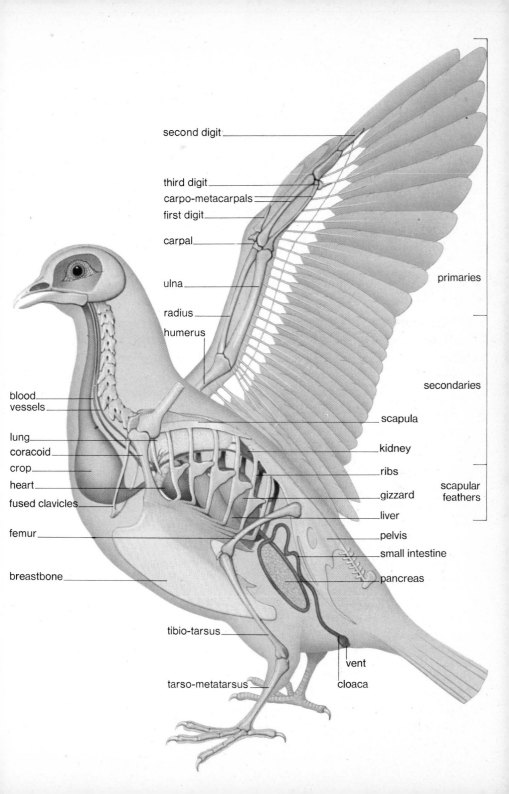

second digit

third digit
carpo-metacarpals
first digit

carpal

ulna

radius

humerus

blood
vessels

lung
coracoid
crop
heart
fused clavicles

femur

breastbone

tibio-tarsus

tarso-metatarsus

primaries

secondaries

scapula

kidney

ribs

gizzard

liver

pelvis

small intestine

pancreas

scapular
feathers

vent

cloaca

Birds are found everywhere throughout the world. Ornithologists believe they number about 100 000 million, more than thirty times the human population. Like other animals, each bird species usually has a particular habitat.

Of the 9000 species more than half belong to the passerine (perching) order, with three toes in front of the foot and one behind. Most passerines are song birds, but even the poor singers have call-notes. Perching birds are land dwellers whose habitats range from tropical jungles to temperate woodlands. There are more than fifty families in this widespread order to which the most common birds belong. The best-known family is the weaver family which contains the common European sparrow and the American song sparrow. Other families contain the finches, buntings, tits, crows, starlings, larks and swallows. Tropical members include the shrikes, umbrella birds and birds of paradise.

Birds of the woodpecker order have two toes facing forwards and two back. Most are tropical

Top: *great spotted woodpecker*. Centre: *yellow bunting*. Below: *Hawaiian jungle fowl*

although some species, such as the California woodpecker and European spotted and green woodpeckers, are temperate. The strangest-looking member is the toucan of South America with a long tongue and an enormous brightly-coloured beak.

Birds of the parrot order, which includes parrots, budgerigars and cockatoos, have a nutcracker beak. The lower bill is moved back and forth against the horny ridges of the upper one to file down hard bits of food.

There are about 300 species of birds of prey. Most of them live together in pairs, and the female is usually larger than the male. This order includes condors and vultures, which feed on dead flesh, bird-hunters such as the peregrine falcon, fish-eaters such as the osprey and bald eagle, and the rodent-eating buzzards and hawks. They are all called diurnal birds of prey because they are most active by day. Owls are nocturnal birds of prey and are placed in a separate order.

The fowl order includes turkeys, pheasants, grouse, partridges and peacocks.

Top: *Australian cockatoo.*
Below: *African fish eagle*

In some areas of the world, especially on isolated islands and open plains with few fast-moving enemies, there are birds that cannot fly. The flightless birds are easy prey for man. The famous dodo, related to the pigeon but bigger than a turkey, flourished on the island of Mauritius until it was exterminated in the 17th century.

The ostriches of the African plains, the rheas of South America and the emus of Australia are all large flat-breasted birds. The breastbone ridge, which is a feature of all birds that fly, is absent in these flightless birds. However, they have large powerful legs for running very fast, and soft fluffy feathers.

Ostriches eat almost anything, but they mainly feed on plants. They can run faster than a horse, with the wings spread out to help lift the body. Ostriches lay 10 or more eggs, weighing about 1·5 kg each. They are the largest living birds, over two metres tall and weighing about twice as much as a man. The wingless moas, which lived in New Zealand until they became extinct a few centuries ago, grew up to twice as tall as an ostrich. The roc of Madagascar may have been

Left: *the ostrich makes a hollow in the ground to lay her eggs.* Top right: *a 17th century painting of a dodo.* Below right: *New Zealand kiwi*

even taller. Its eggs, which are still sometimes found preserved in swamps, have a circumference of almost a metre.

The New Zealand kiwi and the cassowary from New Guinea and Australia inhabit wooded country, where there were few enemies until man arrived. The kiwi is about the size of a large chicken and hunts for worms at night by means of its keen sense of smell and its long thin beak with nostrils at the tip.

Penguins are found only in the cold seas of the southern hemisphere and usually live in large colonies. They eat fishes and crustaceans, and swim very rapidly. Their stiff flipper-like wings are useless for flying but have become specialized for swimming. They usually move about on land in an upright waddle but can also toboggan along the snow on their bellies, pushing with their flippers and legs. Their size ranges from the tiny blue penguin of New Zealand to the emperor penguin, over a metre tall, which breeds not far from the South Pole. The emperor penguin lays a single egg and does not build a nest. The parent incubates the egg on top of its feet in a fold of loose skin where it is protected from the ice on which the bird stands.

Emperor penguins in antarctica

The plumage of a bird consists of many distinctive units called feathers made of a horny substance called keratin. They grow from feather buds in the skin to form a definite pattern over the body.

There are three types of feathers. The flight feathers on the wings and tail consist of a large rigid quill (rachis) growing feathery barbs. The barbs have many tiny hooks that connect with each other so that they form a flat surface (vane). If the feathers tangle, the barbs are 'zipped' together again by combing with the beak. Contour feathers cover the main part of the body like tiles on a roof. Underneath are the down feathers which trap an insulating layer of air between the body and the contour feathers. Because the down feathers have no hooks to connect the barbs, they are feathery instead of forming a flat surface. Newly-hatched birds have only down feathers; the contour and flight feathers develop later. The eider duck lines its nest with great numbers of its fluffy feathers. Masses of these down feathers make eiderdown.

When a feather's growth is complete the quill seals off at the base and the cells making up the feather die. As a bird moults in summer,

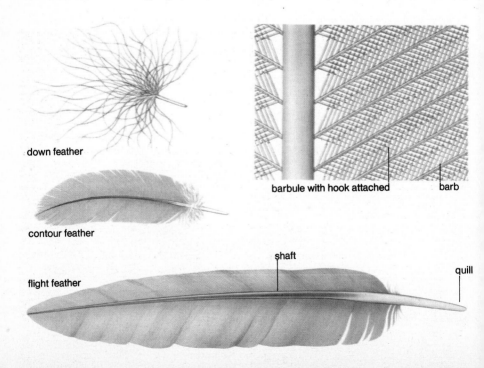

down feather

contour feather

barbule with hook attached   barb

shaft

quill

flight feather

or loses its feathers for any reason, new ones grow from the buds. Birds have no sweat glands, for sweat would probably damage the feathers, but there is an oil gland near the base of the tail. The oil, spread with the beak when preening, keeps the feathers in good condition and prevents the beak from becoming brittle.

The plumage of many birds is speckled or striped, giving a camouflaging coloration which protects them from enemies. Many other birds, especially in the tropics, are brightly coloured. The colours come partly from pigments in the feathers and partly from surface light reflection, just as a film of oil on water may seem brightly coloured. In the latter case the bird's colour will vary depending on the angle from which it is observed. The feathers of the male birds are often more vivid than the female's and sometimes are especially shaped and coloured to attract her. The male peacock has a huge gorgeously-coloured tail which it spreads to attract the female, and the argus pheasant spreads its wings for the same reason. Among the most spectacular are the small birds of paradise of New Guinea which have extraordinarily shaped, brilliantly coloured feathers.

Left: *birds spread oil on their feathers so that water runs off them.* Right: *male peacock*

Birds are adapted for flying in several ways. Their forelimbs are specialized as wings covered with flight feathers; they have powerful wing muscles, a rigid body skeleton, light hollow bones, a large heart and well-developed nervous system. There is also a system of air sacs within the bones and between the body organs which provides extra air for the increased respiration while the bird is in flight.

The wings are concave below and convex above and have a thick front (leading) edge tapering off to a thin (trailing) edge, like the wings of an aeroplane. They provide the initial lift to launch the bird in the air, and then give it forward propulsion through the air. Birds take off with a jump or short run, preferably into the wind, followed by a powerful semicircular beating of wings which produces lift on the downstroke and forward thrust on the upstroke. After gaining height the wings move with an up-and-down flapping, with the lift and thrust coming from the downbeat. The tail helps to steer and the legs are tucked out of the way so that the body is smooth and streamlined.

Waterfowl, such as ducks and swans, have greater difficulty in

Birds in flight. The diagram shows movements of a duck's wings. **Below:** *a spoonbill*

taking off straight from the water without a firm surface to push from. To overcome this they raise themselves with much more wing flapping. When landing, a bird slows down by widening its wings and tail and pushing its body vertically downwards to act as an air brake. A moulting bird which has lost its tail feathers lands badly because it cannot slow down quickly enough. Some birds can fly far and fast. Reliable figures are difficult to obtain, but ducks are said to fly at up to 100 kph and swifts even faster, while the peregrine falcon is said to dive at 150 kph or more. In general, the faster birds have longer and narrower wings. All birds flap their wings when they fly. A humming-bird flaps at up to 60 beats a second, but some birds save their energy by soaring and gliding. Gliding birds such as gulls and albatrosses climb or soar then gradually lose height. Vultures and some eagles soar well, using rising currents of warm air like glider pilots. Gliding birds have long narrow wings, while soarers have long broad ones.

Man's attempts to fly by imitating birds have failed, often fatally. He would need a much bigger heart, and flapping muscles that made up half his body weight, to fly in this manner.

Left: *hummingbirds hover while extracting nectar.* Right: *a tawny owl landing on a tree*

A bird's beak (bill) has several functions: it is used to clean and comb feathers, as a tool for nest-building and as a weapon. It may bear brightly-coloured horny plates as in some pelicans, or fleshy processes, as in those of coots. Most important, the beak, and sometimes the tongue, is used for gathering food. The beak is constantly growing as the old worn-out cells are replaced by new ones.

The bill's shape is usually a clue to a bird's food habits. Warblers, which eat small worms and insects, have a short slender bill. The woodpecker picks out insects from cracks and crevices in trees and logs and has a strong chisel-like beak for hammering into tree bark to reach the insects underneath. Seed-eaters, such as finches and sparrows, have short stout beaks to crack open the seeds. Birds of prey, including eagles, hawks and owls, have sharp-edged beaks for killing their prey and tearing its flesh. Fish-catchers, such as kingfishers and herons, have long dagger-shaped bills with serrated (tooth-like) edges to hold slippery fish.

Some species of woodpeckers can poke their tongue beyond the beak to snatch insects from under bark or ants from the ground.

nightjar

eagle owl

Hummingbirds hover over a flower while they push their small slender bill among the petals and suck up the nectar with their tubular tongue. Parrots have an unusual hooked bill with ridges on the inside for crushing nuts and fruit stones.

A bird's feet are used for running, perching, swimming, fighting and catching and grasping prey. The feet of perching birds have four toes. The first toe is the most powerful and has the longest claw which points to the back of the foot. The other three point to the front. Most birds have only four developed toes; the fifth is merely a little knob at the rear of the foot.

Ostriches are unique in having only two toes; their feet are large and strong and well adapted to running. Ducks, storks, herons, penguins and flamingos have webbed feet, ideal for swimming or walking on soft marshy ground. The large emperor penguin uses its webbed feet as a platform to incubate its single egg and, later, carry its chick. The egg and chick are pressed against the body of the parent for warmth. Birds of prey usually have strong gripping feet with big sharp hooked claws (talons) for grasping and holding their prey.

flamingo   **green woodpecker**

Birds have interesting patterns of behaviour and some have curious ways of feeding. The woodpecker finch of the Galapagos Islands holds a cactus spine in his short bill that helps dig insects from tree bark. A European song thrush feeds on snails, breaking the shells by striking them against a stone. The Egyptian vulture cracks open an ostrich egg by dropping stones on it, and feeds on the contents.

All birds spend a lot of time keeping their feathers in good condition. They bathe themselves in water or sand, dress their feathers with oil from a gland at the base of the tail, and preen by drawing the feathers through the bill to remove ants, dirt and parasites.

There are various behaviour patterns during mating and nesting. Many birds, including the North American sage grouse, gather in special places where the males attract the females with complex dance-like movements called displays.

Some birds nest in a simple hollow in the ground, some, such as the penguins, do not make nests at all, while other birds construct elaborate ones. These may be built either by the hen (female) or cock (male) bird or by both working together. The male weaver bird builds

The male sage grouse, when it is courting, attracts the female with elaborate displays

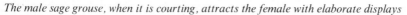

a complicated flask-shaped nest of plant fibres, and the long-tailed tit uses moss, cobwebs and feathers. The blackbird lines its nest with mud, while swallows make theirs entirely of mud. Many other birds, including kingfishers and woodpeckers, dig holes in river banks and trees. If an enemy approaches, the bird either crouches on its nest to protect the eggs and escape attention, or tries to lure the enemy away. Small birds distract birds of prey by mobbing them – flying at them in large numbers with shrill cries. The female birds of some species, such as the blue tit, are fed by their mates while sitting on the eggs.

The young of ground-nesting birds, such as chickens, can run almost as soon as they hatch, but others are usually born blind and helpless. Some young birds have beaks with a brightly-coloured rim which makes their open mouths more obvious at feeding time. Birds that eat small animals give bits of the prey directly to their young. Sea birds and pigeons partly digest their food, then regurgitate it (bring it up) to feed their chicks. Pigeons also feed their young on a milk from the inner lining of the crop.

Top left: *male blue tit feeds a brooding female.* Below left: *robin taking a bath.* Right: *a finch from the Galapagos Islands uses its beak to wield a tool*

The animal kingdom
103   migration: the long journeys

Some bird species regularly travel between one region and another. They use one area during the breeding season and the other for the rest of the year. Migration is mainly from north to south and back again with the change of the seasons. Birds of the northern hemisphere, such as ducks and geese, tend to migrate to northern and near-arctic regions for breeding during the summer months and go south for the winter. Similarly, birds of the southern hemisphere tend to migrate north towards the equator for the winter. Some species, however, are not restricted to a migration range within one hemisphere, but travel across the equator. Swallows and cuckoos breed in Europe and winter in central and southern Africa, while other species migrate between North and South America. Arctic terns probably hold the migration record as they breed in or near the arctic region and winter in the antarctic – a migration range of 17 600 km.

Much of what we know about migratory habits has been learned from ringing. A metal ring recording the date, the country where it was ringed and other details is attached to the leg of a bird. The migratory habits of many species have been recorded in observatories

The map below shows the breeding range and migration routes of the arctic tern

Breeding range   Main migration routes

Winter range

all over the world. Today, migrations are often plotted by radar.

Just before they are due to migrate, birds become very restless and they build up a layer of fat beneath the skin which is used as a source of energy during the journey. It is not known exactly what stimulates the birds to migrate. Migration to warmer places in winter is probably triggered by the shorter autumn days, but there is no obvious trigger for the return journey from tropical regions, which have a constant day length.

Little is known about how birds navigate when migrating. Some birds align themselves to north or south, and keep flying in the same direction until they reach their destination. Other birds, such as homing pigeons, have a more complex navigational mechanism to reach a particular point or area within the country from which they were released. It is thought that recognition and use of landmarks play a role in this type of navigation. We know that young birds do not learn the route from their parents because they often travel alone. In both these kinds of navigation the birds are thought to make use of the sun, moon and stars to orientate themselves.

Left: *chicks have their legs ringed before they migrate.* Right: *arctic tern in flight*

Domesticated four-legged animals such as dogs and cats, and wild animals such as lions and elephants, are all mammals in the same way as man. Their main characteristic is mammary glands in the skin which secrete milk for their young. Mammals are backboned animals (vertebrates), and in most types the young are not hatched from eggs but born alive from within the mother. Unlike all other animals, except birds, mammals are warm-blooded with a natural thermostat to keep body temperature steady at a level at which body processes can work best. Many mammals have an insulating layer of hair which prevents too great a loss of heat but man, whales and seals are much less hairy. Some mammal species have spines or scales instead of hair. All mammals keep cool by losing sweat from the sweat glands in the skin; as the sweat evaporates it takes away the heat. Other features of mammals include an external part to the ear (pinna), different types of teeth to chew, tear, grind or chop food, a well-developed brain, and a muscular sheet (diaphragm) which separates the chest from the abdomen and functions in breathing. The most primitive mammals, the duck-billed platypus and spiny anteater, lay eggs but resemble other mammals in being warm-blooded and suckling their young.

Fossil evidence indicates that mammals developed from small four-legged reptiles called therapsids, which lived about 200 000 000 years ago. Until Cenozoic times, about 65 000 000 years ago, the mammals were small and insignificant, but when the giant reptiles became extinct, mammals became the dominant animals. The marsupials were the first large group of mammals, and they became widely distributed in Tertiary times. Later, however, the early marsupials were displaced almost everywhere by placental mammals, and today marsupials are found only in Australia and parts of the Americas.

Most mammals are land animals, and have become adapted to almost every kind of habitat; they live on plains, in trees, they burrow into the soil and live in jungles and deserts. Some, such as the whales and seals, have fully adapted to marine life; one group, the bats, can fly. Mammals vary in size from the pygmy Etruscan shrew, between five and eight cm long including the tail, to the blue whale, which has grown to a length of over 30 metres and a weight of about 110 tons.

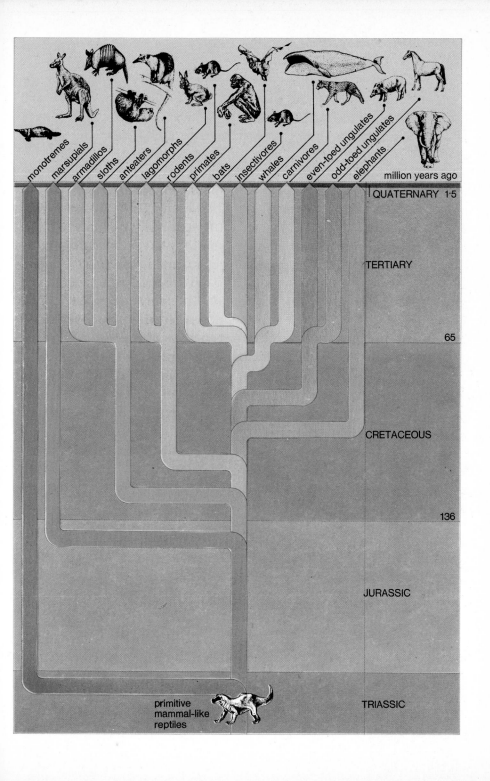

monotremes
marsupials
armadillos
sloths
anteaters
lagomorphs
rodents
primates
bats
insectivores
whales
carnivores
even-toed ungulates
odd-toed ungulates
elephants

million years ago

QUATERNARY 1·5

TERTIARY

65

CRETACEOUS

136

JURASSIC

TRIASSIC

primitive
mammal-like
reptiles

The monotremes, a primitive order of mammals, have many similarities to reptiles. They have only one body opening, the cloaca, from which pass waste matter and eggs or sperm. Their shoulder and hip girdles and eyes have many reptilian features. Like reptiles they lay eggs with leathery shells, and have no external ears. In common with other mammals they have mammary glands that provide milk for the young, but the glands are simple and lack nipples. Monotremes are warm-blooded, although they control their body temperature less well than higher mammals, and they have a body covering of hair. Their hearts and brains are more like mammals' than reptiles'.

There are only two small families in the order of monotremes, both found in or near Australia. One includes the spiny anteater or echidna which has a rounded tailless body, about 60 cm long, covered by a mixture of hair and spines. It escapes enemies by digging a burrow so quickly with its powerful claws that it disappears with great speed. It hibernates through the winter, moults on waking and then grows a new coat. Echidnas feed on ants, grubs and termites with a long beak-like snout and a long sticky tongue on which the ants are trapped.

Duck-billed platypus about to enter the water

Echidnas have no teeth because these are not needed for their kind of diet. The female usually lays only one egg a year which is hatched in a small pouch in her belly. The young anteater sucks milk which trickles from the simple mammary glands of the mother.

The other family contains only the duck-billed platypus. The platypus is about the same size as the spiny anteater, but is covered with short, thick, soft fur. It has a broad flat tail and five-clawed webbed feet, and its mouth, which gives the strange animal its name, resembles a duck's bill, being flat, soft and leathery and rimmed with horny ridges instead of teeth. The platypus is an expert swimmer and diver, using its bill for foraging in the mud for worms, grubs, tadpoles and shrimps which are stored in its cheek pouches until there is enough for a good meal. The horny ridges on the bill are used for straining off mud taken up with the food. The female platypus digs a deep burrow in a river bank in which she lays her eggs. She hatches the eggs like a bird, using the heat from her body. The blind and hairless young do not live in a pouch like young echidnas, but are suckled like other young mammals.

The spiny anteater grows a new coat each summer

Marsupials are a small group of mammals characterized by a pouch on the belly of the female. Unlike other mammals, marsupials are born in a very immature state and immediately crawl into the pouch where they are suckled by the mother and complete their development in the pouch rather like a premature baby in an incubator. Marsupials were found on nearly every continent during the Tertiary age. At that time Australia was still joined to the mainland of Asia, forming one huge land mass. After Australia had split from Asia, the placental mammals competed with marsupials and replaced them in most parts of the world. Because Australia was isolated from the rest of the world the placental mammals never reached there and to this day it is one of the few places where marsupials have spread and increased, unhindered by competition from other mammals.

In Australia there are marsupials that resemble wolves, mice, moles, cats, dogs, bears and other placental mammals. The kangaroo, a mere three cm at

Top: *koalas look very like bears.* Below: *a South Australian wallaby carries her baby*

birth, grows into a large animal which moves with great bounds of up to 10 metres on its powerful hind legs. The wallaby is similar to the kangaroo but smaller. Wombats and koalas look very much like bears. The koala has white-ringed eyes and fluffy fur and lives in trees. Female koalas have only one young at a time, which remains in the pouch for three months and is then carried on its mother's back for another six. It is a timid vegetarian hunted almost to extinction for its beautiful fur. The Tasmanian devil is a ferocious marsupial that has black fur and feeds on flesh.

The only marsupial outside Australia is the opossum, which is common in North and South America. It is about the size of a cat and produces a large litter of young, each no bigger than a man's fingernail. The strongest fasten onto the 12 nipples in the mother's pouch and stay there for several weeks until they are strong enough to cling to her back. 'Playing possum' – deceiving enemies by appearing to be dead – is thought to be a reaction of fear and not a conscious trick.

Top: *Australian brush-tailed phalanger.*
Below: *Tasmanian devil, a ferocious marsupial*

Bats are the only mammals that
can actually fly. They are found
throughout the world except in
polar regions. Zoologists classify
them into two main kinds: the
fruit-eaters and the insect-eaters,
which are usually smaller. Most
bats are active at night; they
usually hunt for food at sunset or
sunrise and rest during the day.
A bat's wing consists of a thin
elastic skin that stretches between the digits of the very long
forelimb, and backwards to the
hind feet. The thumb, with a
hooked claw, is the only digit of
the forelimb that can move separately. Like birds, bats have a big
bony keel on the breastbone for
anchoring the powerful wing
muscles. Fruit bats, such as the
peculiar-looking tube-nosed bat
of south-east Asia and Australia,
are most common in tropical
countries and have the large eyes
and good eyesight characteristic
of nocturnal animals. The smaller
insect-eating bats have poor sight,
although they hunt in the dark,
but often have large ears, as in
the mouse-eared bat. They man-
oeuvre skilfully in flight by sending out supersonic squeaks that

*The tube-nosed bat rests upside down. His out-
size nostrils are an example of specialization*

bounce back from solid objects, including the insects they hunt, and are picked up by their very sensitive ears. This system is very similar to the sonar used for locating submarines. Bats roost in caves, hollow trees or unused buildings, hanging upside down by their claws. In temperate regions they hibernate during the winter months. Most bats mate only once a year, in late summer or autumn, and in most species the female has only one offspring at a time. In hibernating species the young are not born until after the winter hibernation. Young bats are blind and helpless at birth, but are able to fly within two or three weeks. Tropical vampire bats are five to six cm long and have very sharp outer teeth with which they pierce the skin of their victims, most often sleeping cattle. The bats lick up the blood that flows from the cut; they do not actually suck blood. The largest bat is the Malay fruit bat, with a body length of 40 cm and a weight of 900 grams; the smallest is the West African pipistrelle, with a 38-mm body length and a weight of 25 grams.

Top: *mouse-eared bats have large ears*. Below: *the tropical fruit bat lives on fruit*

The small, rather primitive, insectivores are probably closest in structure to the early mammals from which all other types developed. They are mostly nocturnal, feeding on insects, spiders, worms and other small invertebrates, and they have many primitive features, including a simple brain.

Moles are insectivores specialized for burrowing and living underground. They have small eyes and very poor eyesight, but are furnished with large shovel-like front feet with long claws, which make digging easy. Moles are about 15 cm long, and have a smooth streamlined head and body covered with black or brown velvety fur. A mole's nest is an underground chamber lined with leaves, leading to a system of tunnels through which it hunts for food. Sometimes it comes to ground surface at night to hunt for a meal. The presence of moles is usually revealed by the mounds of earth (mole hills) which they throw up while digging their tunnels.

Hedgehogs are larger than moles, growing up to 25 cm long, and are covered with spines on the head and back. They are found

The common mole lives in the gardens and fields of temperate regions. He throws up mounds of earth called mole hills

throughout Europe and Asia, and hunt at night mainly by smell for insects, snails, small snakes and other small animals. Unlike moles they hibernate in winter, curling up in a nest of leaves. When frightened they roll up into a prickly ball, and this makes most of their enemies wary, except badgers which can crunch up anything. Hedgehogs have a very keen sense of hearing.

Shrews, the other main group of insectivores, are small mouse-like animals, rarely seen because they are active at night and move about very quickly, uttering a twittering or screaming cry. They have a long pointed snout, numerous pointed teeth and a very long tail. The smallest living mammal is a shrew; the Etruscan pygmy shrew is only five to eight cm long, including the tail. Small mammals use up energy more quickly than larger ones, and therefore shrews must eat almost constantly to stay alive. They even eat each other when there is no other food. They are very useful to man because much of their food consists of harmful insects. The dark-brown common shrew of Europe and Asia has a body length, not including the tail, of 7·5 cm.

Left: *in spite of his fierce appearance the hedgehog has an unprotected underside.* Right: *the common shrew, active at night, has a longer snout than a mole*

Primates form the order of mammals to which man belongs. Primates other than man mostly inhabit Central and South America, Africa and southern Asia. Although ancestral primates resembled insectivores, later primates developed towards a more active tree-dwelling (arboreal) way of life, with grasping hands and feet, and nails instead of claws. The thumb and big toe are opposable, that is, they can be moved to face the other digits which makes climbing easier. Primates have good eyesight, with both eyes facing forwards to give 3-D vision. Most primates are diurnal – active during the daytime – rather than nocturnal like the insectivores. Their sense of smell is not as good as that of insectivores and other mammals. Primates tend to have a more rounded skull and a relatively larger brain than the other mammals.

The less advanced primates are called prosimians and include tree shrews, lemurs, lorises, bushbabies and tarsiers. Tree shrews are really intermediate between primates and insectivores, resembling shrews in appearance but leading an active arboreal life. Lemurs are found mostly on the island of Madagascar. They have foxy faces and

Left: *bushbabies are an African species. They leap among tree branches at night.*
Right: *slow lorises cannot leap, but move hand-over-hand on branches, often upside down*

long tails and are either diurnal or nocturnal. They feed on fruit and small animals. Flying lemurs are not true lemurs at all, but belong to a separate order, Dermoptera, and live in southern and south-east Asia. They have a fold of skin stretched between the legs which enables them to leap and glide from tree to tree.

Lorises, whose habitat is south-east Asia, are slow-moving nocturnal prosimians. They climb among tree branches, often moving upside down along the underside of a branch. Lorises feed on insects, small birds, lizards and fruit. Bushbabies resemble lorises but have long legs and tail and are fast-moving, leaping acrobatically from tree branch to branch. These nocturnal animals live in African forests, have very large eyes and ears and feed mainly on insects. Some species make good pets. Tarsiers are very strange-looking nocturnal primates that live in trees in the Philippines and neighbouring islands. They have enormous eyes and bat-like ears and long fingers and toes, each ending in an adhesive pad which enables them to climb smooth surfaces.

Flying lemurs glide by taking flying leaps with their limbs outstretched. In this way the skin between the legs acts as a parachute

Monkeys are more man-like than the prosimian primates, with flatter faces and larger, more complex brains. They are basically four-footed but frequently sit upright, using their hands to pick up and examine objects and handle food. Monkeys generally live in family groups or larger herds. This protects them from attacks to which young monkeys are especially vulnerable. Many of their habits are similar to those of primitive human societies; for instance, a powerful dominant male often rules the group like the chief of a tribe. While prosimians are usually nocturnal, most monkeys are active during the day. They are mainly vegetarian, eating fruit, seeds and leaves, but also feed on small animals, especially insects.

Monkeys can be broadly divided into two groups: the New World monkeys of Central and South America, and the Old World monkeys of Africa and tropical Asia which are more closely related to man. The flat-nosed New World monkeys are mainly acrobatic tree-dwellers, and some have a long gripping (prehensile) tail from which they hang, leaving all four limbs free. The tail is also used as a fifth

Left: *the ferocious-looking and dangerous African mandrill is the largest Old World monkey.*
Right: *the smaller golden lion marmoset of South America is an acrobatic climber*

limb when climbing trees, especially by the spider monkey, whose very long, thin arms and legs and small body make it resemble a large hairy spider. The most common South American monkey is the capuchin and the smallest is the pygmy marmoset, measuring only 15 cm long, not including its non-prehensile bushy tail.

Old World monkeys usually live on the ground. They have a well-developed nose and the tail, if present, is non-prehensile. The rhesus monkey belongs to a group called the macaques. It has short limbs and tail and lives in hillside forests, feeding on fruit and insects. The Hindus regard it as sacred and today it is often used in medical research. It begins to breed when it is about four years old, and may live for over 20 years. Another macaque monkey is the Barbary ape of Morocco and Gibraltar, the only monkey now living in Europe. Like true apes, it has no tail. Baboons and mandrills are the largest monkeys, with long dog-like muzzles and large teeth. They live in large troops or family groups and are very fierce, afraid only of attacks from lions.

The long-tailed macaque is human-looking and playful. An Old World monkey, it lives in family groups and is often tamed as a pet

Anthropoid apes are different from monkeys and prosimians because they are completely tailless, with long arms, and a highly-developed brain. They are called anthropoid (man-like) to distinguish them from similar but less advanced primates such as the Barbary apes.

The gibbons are the smallest of the anthropoid apes. They are found in the forests of southern and south-east Asia and are very good climbers, swinging from branch to branch with their long arms. They feed on fruit and nuts, walk on their hind legs when they are not climbing and sleep among the dense foliage of the trees.

The orang-utans, gorillas and chimpanzees are known as the great apes because of their size. Orang-utans are found in the forests of Indonesia, living and feeding among the tree branches. Adult males grow to a height of about one metre and have a large pouch beneath the chin; all orangs have very long arms and are covered with reddish hair. They feed on fruit and vegetables. The other two great apes live in the forests of equatorial Africa.

Top: *gibbons are the smallest anthropoid apes.*
Below: *chimpanzees are friendly and intelligent*

Gorillas have black hair and skin and are the largest living primates, an adult male standing at over 1·5 metres with an arm spread of nearly 2·5 metres. They have reached a body weight of 270 kg. These ferocious-looking gorillas live mostly on the ground in dense forests, travelling about in family groups. They are vegetarians and will not fight unless attacked or annoyed.

The chimpanzees, found in forests and grass-covered regions, are said to be the most intelligent of the apes and the most like man. They show great curiosity, can reason in a simple way, and can even draw and paint complex patterns. In the wild they live in family groups, moving from one feeding place to another and eating fruit, nuts and berries. They communicate with each other in a series of calls and by different facial expressions that resemble the smiles and scowls of humans.

By comparison with the ape, man walks in an upright stance on feet modified for walking rather than gripping. His brain is much larger and his teeth and jaws are smaller.

Top: *the ferocious, forest-dwelling orang-utan.*
Below: *gorillas do not like people*

Rodents, gnawing animals usually of small size, are the largest order of mammals, making up two-fifths of all the mammal species. They all have a single pair of sharp chisel-like incisor teeth at the front of the upper and lower jaws, with a gap between them and the grinding cheek teeth. The gap allows rodents to gnaw with the incisors for long periods without wearing out the cheek teeth. The incisors continually grow to replace the parts worn away by gnawing. Rodents eat mainly plant material and are regarded as pests. The order can be divided into three broad groups: squirrels, rats and porcupines.

The squirrel group includes squirrels, marmots, chipmunks, pocket gophers and beavers. Grey squirrels are native to North America but have been introduced into Britain where, in most places, they have replaced the less destructive red squirrel. They are well adapted to living in trees; they eat young birds, eggs, nuts and berries, and often damage tree bark. Squirrels hibernate in the winter. Beavers are large rodents about one metre long including a broad flat tail. They live in fresh water and swim with their tail and webbed feet. They build large

Left: *the red squirrel lives in woods or wherever there are clumps of trees. It feeds on nuts and berries.* Right: *beavers are always found near water*

houses (lodges) in the water with twigs and branches and use their very sharp teeth to gnaw down trees for constructing dams. The dams may be 20 metres across and enclose an isolated pool with a steady water level in which the lodges are built.

The rat-like rodents include hamsters, lemmings, voles and gerbils, as well as rats and mice. The black rat is found in buildings, sewers and rubbish yards, but has been largely replaced by the bigger, more aggressive, brown rat. Voles are mouse-like rodents that live in the grasslands of Europe and Asia; water voles, or water rats, build complex tunnels along river banks. The house mouse often lives inside buildings and is a serious pest because it eats stored food. The field mouse, on the other hand, very rarely comes near human dwellings. The spiny mouse has sharp prickles instead of soft fur, which protects it from enemies. Lemmings are well known for their migrations, which occur at regular intervals. Vast numbers migrate when their colonies become overcrowded and many accidentally fall into the sea, which has led to the theory that they deliberately kill themselves.

Left: the hair of spiny mice is prickly, making them different from other species. Right: the black rat is a dangerous pest. It caused the historic Black Death and Great Plague in northern Europe

One group of rodents, found mostly in South America, includes porcupines, cavies and guinea pigs, agoutis, coypus, chinchillas and related animals. They are all quite large compared with the other rodents. Old World porcupines live in Africa, southern Asia and southern Europe. They are large burrowing animals, digging deep and complex tunnels in which they live, often in small colonies, and feed on bulbs, roots and bark. American porcupines live in trees rather than burrows, and are expert climbers. The long sharp quills and spines that cover their backs and tails provide an excellent defence against predators. When an enemy approaches the tail quills are rattled as a warning. If the predator advances the porcupine runs quickly backwards and jabs its tail quills into the enemy. The quills of the North American porcupine sometimes become detached during the attack, which has led to the idea that it shoots its spines at its enemies.

Cavies, with round bodies, short legs and no tail, are native to South America. The ancient

Top: *chinchilla, a valuable fur*. Centre: *guinea pig*. Below: *orange-rumped agouti*

Incas bred them for food and also used them for their sacrifices. They were brought to Europe in the 16th century and are the ancestors of the domestic guinea pig. The pig-sized capybara of Central and South America is the largest of the rodents, weighing over 45 kg. It looks somewhat like a small hippopotamus and has partly-webbed toes. Capybaras live near water and feed on the dense vegetation of the banks.

Agoutis are long-legged active South American rodents, about the size of rabbits, that live in tropical forests and feed on fruits, nuts and roots. Their yellow-brown hair is particularly long on the rump where the hairs are raised in a colourful display when the animal is alarmed. The chinchillas are also South American, and are bred in captivity for their silver-grey fur, which is extremely valuable. The coypu is an aquatic rodent, also bred for its fur, known as nutria, which is soft and greyish. Although native to South America it has been successfully introduced into other countries, including Britain where it has become wild.

Top: *prehensile-tailed porcupine*. Centre: *capybara*. Below: *a coypu has soft grey fur*

The animal kingdom
114   rabbits, hares and pikas

At one time zoologists classified rabbits, hares and pikas as rodents because of their superficial resemblance to them. They are now placed in a separate order, the Lagomorpha, which contains between 60 and 70 species. Lagomorphs differ from rodents in bone structure and teeth: they have four upper and four lower incisors, while rodents have only two. The order is widely distributed throughout the world except in antarctic regions.

Rabbits live in underground burrows and have litters of helpless, naked and blind babies. Hares live above ground, making nests called forms, which are simply shallow depressions in the ground. Their young (leverets) are quite well developed when first born; their eyes are open and they are fully furred. All rabbits and hares move fast in leaps, using their long, powerful hind legs. They have long ears, large eyes and a keen sense of smell. When a rabbit senses danger it thumps the ground with its hind feet to warn other rabbits. Hares are solitary and when they sense danger they instantly freeze and then, at the last moment, zig-zag off at great speed. The jack-rabbit of

Left: *barren mountainous regions are home to the Rocky Mountain pika.*
Right: *the rock hyrax, very common in Africa, lives in large groups and seems unafraid of humans*

North and Central America is one of the largest American hares. In 1859 a few rabbits were introduced from Europe into Australia. Because they had no natural enemies there, they multiplied and spread rapidly, causing great losses of crops to farmers. Efforts to check their numbers by using foxes and the disease myxomatosis have not had any long-term success.

Pikas live in the cold mountainous areas of North America and Asia. They are smaller and rounder than rabbits and hares, with short broad ears, short front and back legs and brown fur. They often make their burrows under outcrops of rock, and live in colonies that may number several hundred. Towards the end of summer they make haystacks of dried grass to provide them with winter food.

Another group of animals that resemble rodents are the hyraxes, also called dassies, found mostly in Africa. Their closest relatives, however, are thought to be the elephants. They are about the size of rabbits but have strange hoof-like toes which enable them to climb nimbly among rocks and trees.

White-tailed jack-rabbits, like several other hare species found in snowy climates, are effectively camouflaged in winter. Their coats are greyish-brown for the rest of the year

There are three completely different groups of animals, all called anteaters. The spiny anteater (echidna) is one of the primitive egg-laying mammals. The scaly anteater (pangolin) is found in tropical Africa and Asia, and the true anteater belongs to the order Edentata and lives in South America. One of the edentate anteaters is the South American giant anteater, about the size of a very large dog, which has a long tubular snout and no teeth and a long bushy tail. It has grey-brown fur with dark blotches, short thick legs, and powerful digging claws used to rip open the nests of termites on which it feeds. Its very long sticky tongue catches the termites.

Pangolins are covered on the head and back with sharp-edged overlapping bony plates and look like giant walking pine cones. They are toothless, have a long sticky tongue and powerful digging claws, and curl up for protection. The termite-eating African aardvarks are thought to be unrelated to any other anteater. They have a pig-sized body and large ears and live in burrows.

Top: *two-toed sloth.* Below: *aardvarks are unrelated to other mammals*

They are nocturnal and therefore seldom seen by man.

There are two other edentate families, the armadillos and the sloths, both found in South and Central America. Despite their name (edentate means toothless) they do have teeth, but only in the cheek region. Armadillos live in grasslands and semi-deserts and feed on insects, grubs, earthworms and the fruits of some plants. Their main feature is a shell of bony plates arranged in overlapping bands on the back. Some can roll themselves into a ball. The nocturnal armadillos have powerful digging claws to excavate their burrows. Their keen sense of smell locates food in the soil. Sloths are slow-moving animals that spend most of their time hanging upside down in tropical forest trees, feeding on leaves. They rarely climb down to the ground where they have difficulty in walking. There are two main kinds of sloths, the three-toed and two-toed. Until a few thousand years ago giant bear-like ground sloths, about the size of small elephants, lived in South America.

Top: *scaly anteaters.* Centre: *great anteater, related to the sloth.* Below: *armadillo*

There are three orders of marine mammals: seals (Pinnipedia), whales and dolphins (Cetacea) and manatees and dugongs (Sirenia).

The order Pinnipedia contains the seals, sea-lions and walruses, whose limbs are specialized as flippers for swimming. They also spend much time on shore, especially in the breeding season, using their flippers to drag themselves along the ground. They are carnivorous (flesh-eating), feeding on fish and shellfish. The performing seals in circuses are actually sea-lions and differ from true seals in having external ears and a greater skill in moving about on land. True seals are hunted for their fur, flesh and oil. The elephant seal is the largest, and grows to a length of seven metres and a weight of four tons. Walruses are characterized by long tusk-like canine teeth with which they heave themselves out of the water and catch the molluscs on which they feed.

Whales are more completely adapted to life in the sea than seals, and cannot move about on land. They have only front limbs, which are specialized as small steering flippers. Many have a dorsal (back) fin and a long horizontally-flattened tail for swimming. The nostril

common dolphin 2.60m.

grey seal 3.30m.

opening, called the blow-hole, is at the top of the head and when cetaceans surface after being submerged (for up to an hour in some cases) they blow out air from their lungs as a spout of water vapour. They have a nearly hairless body with a thick layer of insulating blubber (fat) beneath the skin. The young are born and suckled at the surface of the water.

The largest whales are toothless, with sheets of horny material called whalebone inside their mouths, which filters off the plankton on which they feed. The largest of the whalebone whales is the blue whale, also the largest of all mammals, which is about 26 metres long and weighs well over 100 tons. These huge animals are now becoming rare because so many of them have been hunted for their blubber and flesh. The smaller, toothed whales include the porpoises and dolphins. They feed on fish and other sea animals. Dolphins have a more pointed snout than porpoises, but both appear to have evolved a language of squeaks and similar noises. The related black-and-white killer whales (grampuses) grow up to 10 metres long and hunt in packs, killing other whales, seals and birds.

blue whale 26m.

Carnivores are a large group of mainly flesh-eating mammals belonging to the order Carnivora. They all have strong jaws and large sharp teeth specialized for tearing and cutting flesh. Many have powerful claws used to attack enemies and prey.

The dog family (Canidae), found throughout the world, is one of the largest groups of carnivores. It includes wolves, jackals and foxes, as well as dogs, all of whom communicate by means of barks, howls and snarls. The wolf, jackal and domestic dog all belong to the same genus (*Canis*), and can usually breed with one another. The domestic dog is probably descended from a race of wolves. In spite of the large number of breeds, all looking very different from each other, all dogs belong to a single species, *Canis familiaris*. Wolves live in packs and freely roamed the forests of the northern hemisphere but are now much less common because they have been hunted so extensively by man. The pack is usually a number of small family groups of four or five animals with a leader or dominant wolf. Jackals live in the grasslands of Africa and central Asia and hunt alone or in packs. They feed on carrion but also hunt living animals. The coyote is similar to the jackal but lives in Central and North America. The jackal and coyote are both smaller than the wolves. Foxes, unlike dogs and wolves, do not hunt in packs. They live in most parts of the world, from arctic regions of the north (the arctic fox) to the African desert (the tiny fennec fox).

All bears belong to the same family (Ursidae); they are heavily built and walk on the soles of their feet. Bears have large claws, small ears and eyes and most of them eat fruit, nuts and honey as well as flesh. Polar bears of the arctic regions feed mostly on seals and fishes. They are also excellent swimmers. The brown bear, called the grizzly in North America, is widely distributed in northern forests.

A carnivore that resembles the bears but is more closely related to the raccoons is the giant panda, one of the few plant-eating carnivores. Its habitat is the bamboo forests of China, where it feeds mainly on young bamboo shoots, but several specimens have thrived in zoos.

An animal killed by a predator who ate his fill, also makes a meal for scavengers. The vulture will not attack the feeding jackal because both animals eat only dead flesh

The Mustelidae is a large family of carnivores that includes weasels, polecats, badgers, otters and skunks. There are also valuable fur-bearing animals such as minks, sables and martens. Most of this family have long bodies and tails and short legs.

The weasel has a bright reddish-brown coat with a white belly. It preys on small rodents, reptiles, frogs and birds and is found in Europe, Asia and northern Africa. The stoat is similar to the weasel, apart from its black-tipped tail. Those living in northern regions grow a white winter coat, which provides good camouflage in the snow but is also the source of ermine.

Minks occur in the wild in North America, Europe and Asia, but are also bred on fur farms for their extremely valuable pelts. Sometimes they escape from the farms, start living in the wild and become pests.

Badgers, with a thickset body and powerful clawed forelegs, dig complex burrows (sets). They are secretive in their habits, and have therefore survived in well-populated areas where they feed

Top: *leopard*. Top centre: *weasel*. Lower: *silver-haired feral mink*. Below: *badgers*

on rabbits, rodents, insects, worms, berries, nuts and acorns.

In the cat family (Felidae) are the lion, tiger, leopard, jaguar and cheetah, as well as the smaller lynx, bobcat, ocelot and domestic and wild cats. They are characterized by claws which can be sheathed. Most cats, even the big ones such as lions and tigers, can purr. The domestic cat and most of the smaller cats belong to the genus *Felis*. The wild cat of northern Africa is thought to be the ancestor of the domestic cat. Many of the smaller cats, such as the lynx and golden cat, are beautifully marked or coloured.

Today lions flourish in the wild only in central Africa. They are the only social cats, living in groups (prides) in which there are usually several adult males. Lions hunt and kill only when they are hungry, feeding about once a week. The leopard has a distinctive spotted coat and is native to most of Africa and southern Asia. It is an expert climber and lives in forests. The jaguar is similar in appearance to the leopard, but lives in the forests of tropical and subtropical America.

Lions are usually lazy and content until they are hungry; then they become dangerous

Millions of years ago the elephant order was very large; there were many types, both large and small, and even some with four tusks. As late as the Stone age, great woolly mammoths with immense curved tusks roamed Europe, Asia and North America. Today the order is reduced to only two species, the African and Indian elephants.

African elephants have bigger ears and tusks than the smaller Indian elephants; they also have a flatter forehead and a concave back. Their tusks develop from incisor teeth in the upper jaw which grow up to 3·5 metres long in males but are much smaller in females. So many African elephants have been hunted for their valuable ivory tusks that the species is now quite rare and most of the survivors are in game reserves. The Indian elephant has been domesticated for centuries to carry large loads and perform other heavy work. Only the male Indian elephants bear tusks, and these are used for defence, to uproot trees and to dig for roots and water.

Elephants are the largest of all land animals. The African bush elephant is the greatest in size, with males growing up to 3·5 metres tall and weighing up to six tons. Elephants have just one enormous

Left: *the African elephant.* Right: *Indian elephants drink 30–50 gallons of water every day*

ridged molar (cheek tooth) in each half of each jaw, which is used for crushing food. When a molar wears out it is replaced by another one from behind it. An elephant is mature at 15 years and usually dies at the age of 60. Cow elephants produce one offspring every two years. The young calf takes 21 months to develop inside the mother and then takes five years to be fully weaned. The gestation period is the longest of all mammals in comparison with nine months in man and three weeks or less in mice.

The elephant's trunk is an elongated nose (proboscis). It is prehensile (gripping) and is used for gathering food, often up to 100 kg per day of leaves, tree bark, grass and roots. Its delicate tip, which has one finger-like projection in the Indian, and two in the African elephant, can pick up small objects. The elephant drinks by sucking up water with its trunk which is then squirted into its mouth. It can give itself a shower-bath of water or mud in the same way. Although elephants have poor sight they have a keen sense of smell. When they move their trunks from side to side or raise them in the air they are smelling out their surroundings.

African elephants like wading into water to bathe and drink. Both sexes have long tusks

The two orders of hoofed animals (ungulates) are Perissodactyla (odd-toed ungulates) and Artiodactyla (even-toed ungulates). Ungulates are plant-eating (herbivorous) animals, mostly unable to defend themselves, but many have very long legs and can run or gallop very quickly to escape their enemies. The body weight of the odd-toed ungulates is supported by the middle toe of each foot. They do not chew the cud and have no true horns or antlers. There are three families of odd-toed ungulates: horses, tapirs and rhinoceroses. Horses have a single middle toe on each foot, rhinos have three toes on each foot and tapirs have four toes on the front feet and three on the hind feet.

The horse family (Equidae) includes horses, zebras and asses. Man first domesticated the horse to draw carts and chariots; only much later did he ride it. Today horses are used for racing, riding, pulling carts and for all sorts of work in farming and forestry. All our modern horses are thought to have developed from a single wild type. The only wild horse now surviving is Przewalski's horse, which lives in the Mongolian Gobi Desert. It resembles the domestic horse but has

Left: *the horse has a single-toed hoof.* Right: *the white rhinoceros has three toes*

a stiff mane of hair. The zebra, found on open bushland in Africa, has not been domesticated, but the wild ass from desert and mountainous areas of Africa and Asia was probably tamed before the horse. It is larger and more agile than domesticated asses and is a plain greyish-brown colour. The domestic donkey is descended from the wild ass of north-east Africa.

Tapirs, which live only in the dense forests of Central America and Malaysia are shy nocturnal animals. They are stockily built, with a broad down-turned movable snout, simple teeth and short slender legs.

Rhinoceroses are found in the grasslands and bush of Africa and southern and south-east Asia. They are massive clumsy-looking animals covered with a very tough armour-like skin. The horn of a rhinoceros is only a mass of horny fibres; it contains no bone as do the horns of sheep and cattle. The black rhinoceros of Africa is the most common species. The white rhino, also of Africa, is much larger and heavier than the black rhino, and is rated as the second largest land mammal after the elephant.

The coat of a young tapir has stripes which disappear as it grows older

There are nearly 200 species in the order Artiodactyla (even-toed ungulates). Their body weight is carried on the third and fourth toes of each foot, producing the cloven hoof typical of the order. Even-toed ungulates have a complicated digestive system, with a two- to four-part stomach. Most of them are ruminants with a four-part stomach and the habit of chewing the cud (partly digested vegetation regurgitated from the first two stomachs and thoroughly chewed in the mouth before passing to the other two). All even-toed ungulates have bony outgrowths on the top of the head: horns in cattle and antelope and antlers in deer.

The chief non-ruminants are the pig and hippopotamus, with prominent canine teeth which grow into tusks. There are various types of wild pig as well as the well-known domestic varieties. Wild pigs feed mostly at dusk; they root in the ground for plant matter, worms and insects. True pigs, native to the Old World, include the wild boars, which are now most common in Asia, and the warthogs, bushpigs and forest

Even-toed ungulates. Top: *reticulated giraffe.* Below: *hippopotamus*

hogs of Africa. The peccaries of Central and South America are the New World relatives of the true pigs.

There are only two species of hippopotamus, both of which are African. The common hippopotamus weighs up to four tons and spends a great deal of time in the water, where it can remain submerged for up to five minutes; at night it makes its way into the forest to feed. The much smaller pygmy hippopotamus has longer legs and a shorter body than the common hippo and weighs only about 880 kg.

Among the ruminants, antelopes, such as the impala and gazelle, are the most graceful with their long legs for quick escape from lions and other predators. Deer have solid bony antlers which, unlike horns, are grown and shed each year. Also in the deer family are the reindeer and elk, called caribou and moose in North America, as well as the European red deer. The gawky giraffe lives in Africa. A less well-known relative of the giraffe is the okapi, with a shorter neck and a brown-and-white striped body.

Top: *a red deer sheds its antlers every year*.
Below: *the impala has particularly long legs*

Many of the ruminants belong to the cattle family (Bovidae), including oxen, bison, buffaloes, antelopes, sheep and goats. They all have horns that grow from a bony core on the skull. Sheep were among the first animals to be domesticated; their wild ancestors were probably the mouflons of Asia, Sardinia and Corsica. Different types of sheep have been bred for their meat, wool and milk; in the Middle Ages they provided much of England's wealth through an extensive trade in wool and cloth. More recently sheep were introduced into the Americas, Australia and New Zealand. Goats are hardy animals of wild mountainous regions. Domesticated goats are kept in less fertile lands for their milk, meat and skin, but they have often reduced the land to a near-desert state by their constant grazing.

All over the world various species of cattle have been either hunted or domesticated by man. This has led to some species being almost or entirely exterminated while others, such as domestic cattle, are bred in many countries for their meat and milk. The American bison, wrongly called the buffalo, and the European bison have suffered badly. The European bison is now extinct in the wild; today the few

Top left: *the African buffalo was once hunted for game.* Below left: *wild Asian buffaloes*

survivors are carefully protected in zoos and parks. Even more tragic was the fate of the American bison. In 1800 there were more than 60 000 000 roaming the prairies but after a century of reckless slaughter for their meat and hides only about 500 survived. They have been saved from extinction and there are now several thousand in parks and reserves. The aurochs of European forests were probably the ancestors of modern domestic breeds of cattle but they have now completely died out. Wild Asian buffaloes, often called water buffaloes, live near rivers and like wallowing in mud; they are bred in southern Asia for their milk and used like oxen for pulling carts and ploughs. The African buffalo is much less common now than it was a hundred years ago. It was hunted for game but is now mostly confined to reserves. In Tibet the yak is used for riding, as a beast of burden and for its milk and meat. Today there are probably about 800 000 000 domestic cattle throughout the world. Of these, some 180 000 000 are in India, but since the cow is sacred to the Hindus, it is not a source of meat and is used only for ploughing and for milk.

The Indians of the North American plains depended on the meat and hides of the American bison, or buffalo, for their survival. Today the animal is almost extinct

Ecology is the study of plants and animals in their natural surroundings. The world can be divided into a number of regions, each
supporting a particular type of wild life. These plants and animals are
adapted to live in the particular conditions of their region, which are
usually determined by climate and rainfall. This is why some plants
and animals are found in certain parts of the world but not in others.
Ecologists call these regions biomes, and each biome usually takes its
name from the dominant, or most common, kind of vegetation which
determines all other plant and animal life found in that biome.

The temperate forest biome, characterized by deciduous trees, is a
common feature of regions where summers are warm and winters
cold. The dominant trees, which shed their leaves every year, include
the oak, beech and maple.

The coniferous forest biome is found in northern regions where
temperatures are low throughout the year. The dominant plants are
the spruce and pine and the most common animal is the deer.

Even further north is the cold and windy tundra biome, where the
most common plants are lichens, mosses and small shrubs. The

biomes of the world

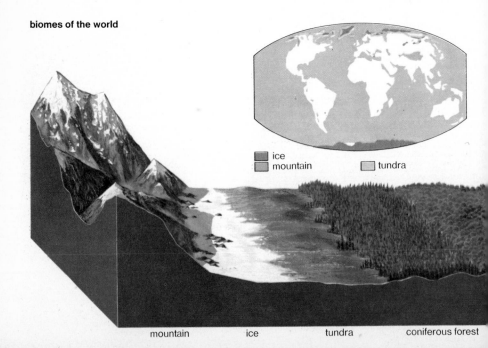

ice
mountain   tundra

mountain   ice   tundra   coniferous forest

animals living there, such as the musk ox with its long shaggy coat, are adapted to survive the intensely cold winters.

At the centre of many continents is the grassland biome, dominated by low herbaceous plants, especially grasses. It supports bison and other grazing animals and burrowers such as the prairie marmot and ground squirrel.

The savannah biome is similar to the grassland, but has more trees; animals living there include the zebra and giraffe.

The desert biome supports only plants and animals adapted to store water and resist drought. The most common desert plant is the cactus, while desert animals include the camel and the scorpion.

The tropical forest biome is characterized by heavy rainfall and high temperature. It supports an enormous variety of plants, dominated by broad-leaved evergreen trees, and many exotic types of birds, insects and mammals. Mountains in tropical regions can themselves be divided into biomes. Tropical forest at the bottom grades into coniferous and deciduous forests and tundra higher up, as the climate becomes colder.

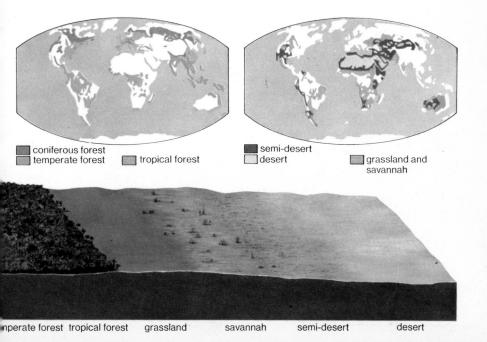

coniferous forest
temperate forest   tropical forest

semi-desert
desert   grassland and savannah

nperate forest tropical forest   grassland   savannah   semi-desert   desert

Plants and animals in natural communities depend on one another and on their non-living surroundings for food and therefore life. All plants and animals must have four elements: carbon, nitrogen, hydrogen and oxygen. These elements are combined to form proteins, fats and carbohydrates in the animal or plant, and are then used for building cells and tissues or as a source of energy. Oxygen is obtained from the air and also from water, in which it is dissolved. It is produced by plants during photosynthesis and used up during respiration, so that the amount present is always constant. Water is composed of hydrogen and oxygen and thus is a source of hydrogen for living things. Carbon and nitrogen come from the air, sea or soil, and the use of these elements by animals and plants involves complex cycles which demonstrate the dependence of animals and plants on each other and the balance existing between them.

In the nitrogen cycle of land-living organisms, nitrogen compounds (mostly nitrates) are absorbed from the soil by plants and built up into proteins and other complex constituents of plant tissues. The plants may be eaten by animals and the animals and plants eventually

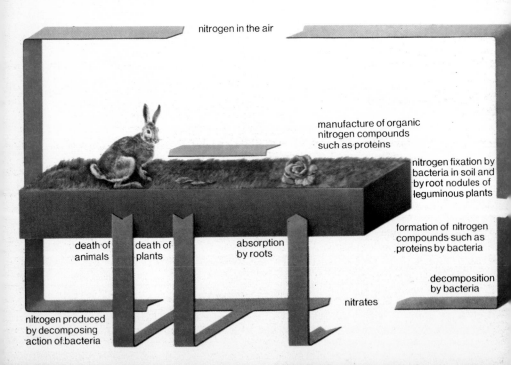

nitrogen in the air

manufacture of organic
nitrogen compounds
such as proteins

nitrogen fixation by
bacteria in soil and
by root nodules of
leguminous plants

death of
animals

death of
plants

absorption
by roots

formation of nitrogen
compounds such as
proteins by bacteria

decomposition
by bacteria

nitrates

nitrogen produced
by decomposing
action of bacteria

die. Bacteria in the soil cause decomposition of the dead animals and plants and during this process the tissues are converted to nitrates, which are thus returned to the soil.

During decomposition some free nitrogen is produced, and this forms part of the pool of nitrogen present in the atmosphere. Certain bacteria in the soil and in the root nodules of peas, beans and other leguminous plants can use atmospheric nitrogen to build up their tissues: this process is called nitrogen fixation. However, most organisms can only use nitrogen in the form of nitrates or more complex nitrogen compounds.

In the carbon cycle, carbon dioxide from the atmosphere is used by plants during photosynthesis to build up their tissues. Some plants may be eaten by animals. Eventually the animals and uneaten plants die and are decomposed by the action of bacteria. Plants, animals and bacteria all give out carbon dioxide during respiration which keeps the amount of carbon dioxide in the atmosphere at a steady level.

There are similar carbon and nitrogen cycles among the animals and plants living in seas and rivers.

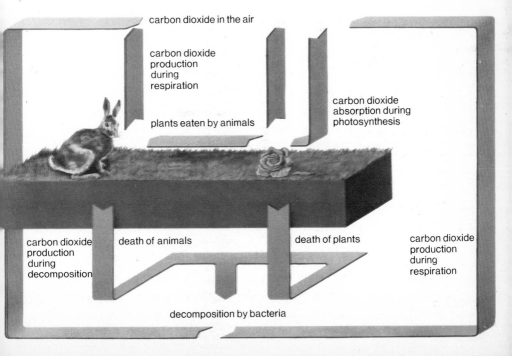

carbon dioxide in the air

carbon dioxide production during respiration

carbon dioxide absorption during photosynthesis

plants eaten by animals

carbon dioxide production during decomposition

death of animals

death of plants

carbon dioxide production during respiration

decomposition by bacteria

All the different plants and animals in a natural community are in a state of balance. This balance is achieved by the plants and animals interacting with each other and with their non-living surroundings, as was shown in the nitrogen and carbon cycles in the previous pages. An example of a natural community is a woodland, and a woodland is usually dominated by a particular species of plant, such as the oak tree in an oak wood. The oak tree in this example is therefore called the dominant species but there are also many other types of plants, from brambles, bushes and small trees to mosses, lichens and algae growing on tree trunks and rocks.

The plants of a community are the producers: they use carbon dioxide, oxygen, water and nitrogen to build up their tissues using energy in the form of sunlight. The plant tissues form food for the plant-eating animals (herbivores) which are in turn eaten by the flesh-eating animals (carnivores). Thus plants produce the basic food supply for all the animals of a community. The animals themselves are the consumers, and are either herbivores or carnivores.

Examples of herbivores in a woodland community are rabbits, deer, mice and snails, and insects such as aphids and caterpillars. The herbivores are sometimes eaten by the carnivores. Woodland carnivores are of all sizes, from insects such as beetles and lacewings to animals such as owls, shrews and foxes. Some carnivores feed on herbivores, some feed on the smaller carnivores, while some feed on both: a tawny owl will eat beetles and shrews as well as voles and mice. These food relationships between the different members of the community are known as food chains or food webs. All food chains start with plants. The links of the chain are formed by the herbivores that eat the plants and the carnivores that feed on the herbivores. There are more organisms at the base of a food chain than at the top; for example, there are many more green plants than carnivores in a community.

Another important section of the community is made up of the decomposers. They include the bacteria and fungi that live in the soil and feed on dead animals and plants. By doing this they break down the tissues of the dead organisms and release mineral salts into the soil.

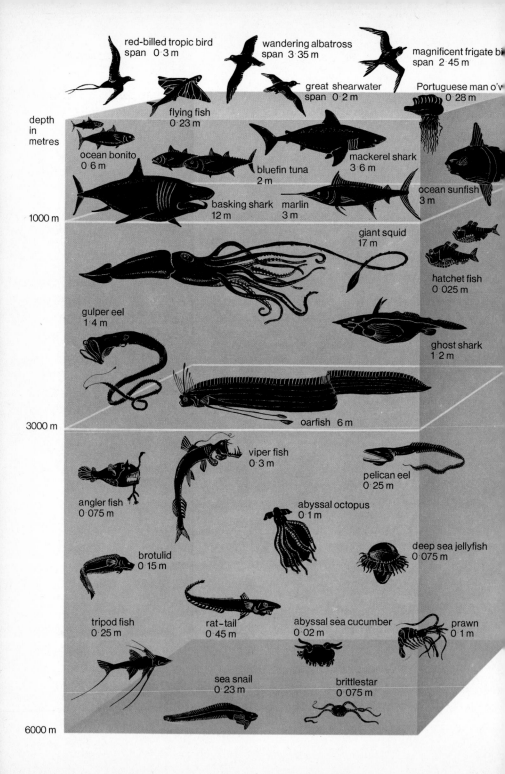

red-billed tropic bird
span 0·3 m

wandering albatross
span 3·35 m

magnificent frigate bird
span 2·45 m

great shearwater
span 0·2 m

Portuguese man o'war
0·28 m

flying fish
0·23 m

depth
in
metres

ocean bonito
0·6 m

bluefin tuna
2 m

mackerel shark
3·6 m

ocean sunfish
3 m

basking shark
12 m

marlin
3 m

1000 m

giant squid
17 m

hatchet fish
0·025 m

gulper eel
1·4 m

ghost shark
1·2 m

oarfish 6 m

3000 m

viper fish
0·3 m

pelican eel
0·25 m

angler fish
0·075 m

abyssal octopus
0·1 m

deep sea jellyfish
0·075 m

brotulid
0·15 m

tripod fish
0·25 m

rat-tail
0·45 m

abyssal sea cucumber
0·02 m

prawn
0·1 m

sea snail
0·23 m

brittlestar
0·075 m

6000 m

The open sea, which also includes the air above the surface of the water, is a natural community. It may be divided into various layers, each with its own characteristics, and supporting different types of animal and plant life. In the air above the surface of the sea are sea birds such as the wandering albatross and frigate bird. The wide span of their wings allows them to hover and soar in between darting down to catch fishes on the surface of the water.

At the sea surface there may be flying fishes and a large floating coelenterate (related to the jellyfish) called the Portuguese man o' war. Seaweeds and phytoplankton – a floating mass of microscopic algae and diatoms – are in the top layer of sea. The plants are confined to the top layer because the sunlight necessary for them to carry out photosynthesis only penetrates to a certain depth of water. As in a woodland community, plants form the basis of the food supply for most of the animals living in the sea.

Down to a depth of 3000 metres below the surface, there are many free-swimming fishes, such as the bonito, sunfish and bluefin tuna and cephalopod molluscs such as the squids. Many of these animals may migrate to the lower levels.

In the lower layers of the sea there are fewer animals, and they tend to eat each other because there is no plant life. The gulper eel, which is not a true eel, has an enormous mouth which engulfs prey much larger than the fish itself. Other fishes at this depth are the ribbon-like oarfish which has very long, thin oar-like ventral fins, and the hatchet fish. This small silvery fish resembles a miniature hatchet in appearance and has a number of light-producing organs at the lower edge of the body. It is often eaten by tuna fish.

At the very bottom of the sea no light penetrates and the animals are usually of a dark colour. They are adapted to withstand the intense cold and tremendous pressure and are often grotesque in appearance, such as the angler, rat-tail and tripod fishes. They are scavengers, feeding on anything that falls from above, or predators, such as the angler fish. Some rat-tails have very long snouts to dig for food on the sea-floor. Other animals found near the sea-bottom are sea cucumbers, brittlestars and crustaceans such as prawns.

The world-wide population explosion in recent years has led to a corresponding increase in the number of factories, fisheries, farms and other industries needed to support it. All these industries provide more food and resources for people but they also produce more waste products which are often harmful to plants and animals and upset the balance of their communities. The term pollution is used for any process that spoils any part of the environment.

One of the most serious problems is pollution of the atmosphere by smoke, dust and chemicals, such as sulphur dioxide. Although this type of pollution is more common in industrial areas, it can harm the wild life of the surrounding countryside. Another source of air pollution is the lead which is added to petrol to increase its efficiency but which contaminates vegetation close to busy roads. A more serious type of pollution for wild life, however, is the contamination of rivers and lakes by the waste products of industry discharged into them. Industrial sewage and the waste products of farm animals use up the oxygen in the water so that algae, fishes and other inhabitants cannot survive. Another way of polluting rivers and lakes is by increasing their temperature and this is likely to occur near a power station which uses water to cool its machinery. The plants and animals living in the water can only survive within a certain temperature range; if it exceeds this limit they will die. Crude oil released from tankers also causes the death of molluscs, crustaceans and other sea animals and sea birds, when their feathers become oil-clogged.

The use of chemical weedkillers and pesticides (poisons used to kill insect pests) causes pollution of the countryside. One of the best-known pesticides is D.D.T. In small doses it does not harm animals. However, it does tend to accumulate in their tissues so that it eventually either poisons them or the animals that feed on them. For example, birds feeding on insects and earthworms contaminated with D.D.T. in the soil are soon poisoned themselves. Another, more poisonous pesticide, is dieldrin. This is used to control the soil pests of crops, but has also killed the birds that feed on the crop seeds and the mammals that hunt the birds. Alternative methods of pest control are under investigation to avoid the use of these dangerous chemicals.

The human body can be divided into a number of separate systems, each with a special job. The skeleton (bony system) provides a supporting framework and protects delicate organs such as the brain and heart. It also works together with the muscular system to provide a means of moving about. The breathing, circulatory and digestive systems co-operate to set free the energy and provide the materials needed for growth and repair of the body. The urinary system disposes of some of the waste produced by the chemical activities of the body. The reproductive system enables babies to be conceived, fed and protected before birth.

The endocrine system is a number of glands, such as the thyroid, sex and adrenal glands that secrete chemicals called hormones into the blood. These hormones control growth, sexual activities, some aspects of food digestion and many other body functions. The nervous system keeps all the other systems under control and enables human beings to think, reason and make decisions. The separate systems all work together for the proper functioning of the body.

Each system has its own organs. The lungs, for instance, are organs in the breathing system, the heart is an organ in the circulatory system and the liver is an organ in the digestive system. The organs in the human body are similar to those in other mammals; they serve the same sort of systems and they are arranged in a similar way.

Every organ is made up of several different kinds of tissues. Epithelial tissue, which includes the skin, forms a covering over organs, and connective tissue includes bone and cartilage. Other types of tissues are muscle, nerve and blood. Every piece of tissue is made up of units called cells, and each different type of tissue is composed of similar cells. Most cells are so small that they have to be measured in thousandths of a millimetre and can only be seen through a microscope. There are more than 50 million million cells in a human body. The largest is the human egg, which is about as big as the head of a pin. Each cell is covered by a thin membrane enclosing a nucleus and a jelly-like substance (cytoplasm) which in turn contains tiny particles, each with their own particular function.

An 18th century manuscript on anatomy compares the human body structure with that of a house

The bones connect to form the skeleton, which provides a supporting framework and an anchorage for the muscles which move the body. It also protects vital body organs, such as the brain, heart and lungs. It weighs about one-fifth of a man's total weight and is made up of over 200 bones. The longest is the thigh bone and the smallest are the three tiny bones in the ear. The most important bones of the skeleton are the backbone, skull, rib cage, and the shoulder girdle and hip girdle, to which are attached the bones of the arms and legs.

Bones are joined together in various ways. The point at which two bones meet is called a joint. Some joints, such as those in the skull, are firmly fused together but most joints are movable. In the movable joints the ends of the bones are covered with smooth gristle (cartilage) and many of these joints are enclosed in a wrapping filled with a watery lubricating fluid (synovial fluid). The bones are joined together by unstretchable cords called ligaments.

The arm and leg bone structure is similar. There is a single upper bone (femur in the leg, humerus in the arm), connected by a ball-and-socket joint to the hip (in the leg) and shoulder (in the arm), so that the limb swivels easily. At the knee the femur is connected by a hinge joint to the two bones of the lower leg, the tibia and fibula. In the same way the humerus of the arm is connected at the elbow to the radius and ulna. Man is unusual among mammals in that the radius can swivel around the ulna, turning the palm of the hand up and down. The hands and feet each contain more than 20 small bones.

Bone is living tissue. The skeleton of a newly-born baby is mainly cartilage, but most of this is gradually replaced by bone as the cartilage cells are impregnated with hard calcium compounds. The living cells in bone are continually being replaced, which enables broken bones to mend easily. Some cartilage never changes to bone, for example, there is always cartilage at joints, in the end of the nose and in the external part of the ear.

Most bones have a spongy marrow in the centre. In many, such as the ribs, hips and backbone, the marrow is red due to a plentiful blood supply and this is where the red blood cells are made. The hollow shafts of the long limb bones contain fatty yellow marrow.

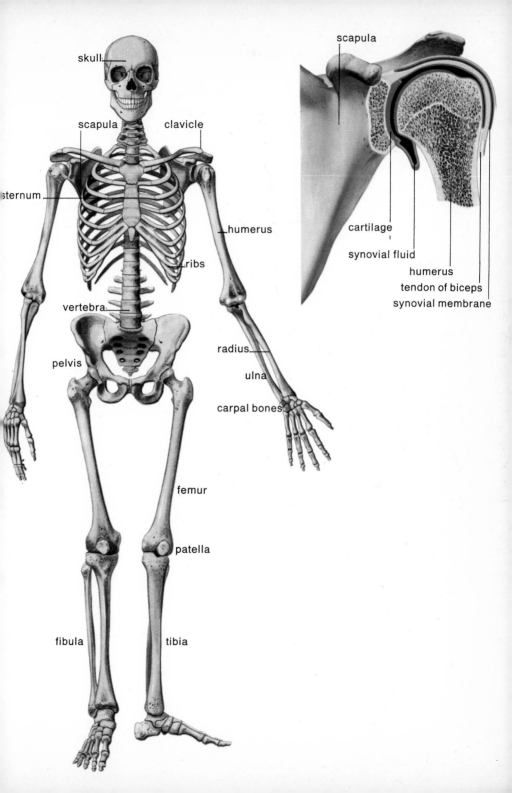

skull

scapula

clavicle

sternum

humerus

ribs

vertebra

pelvis

radius

ulna

carpal bones

femur

patella

fibula

tibia

scapula

scapula

cartilage

synovial fluid

humerus

tendon of biceps

synovial membrane

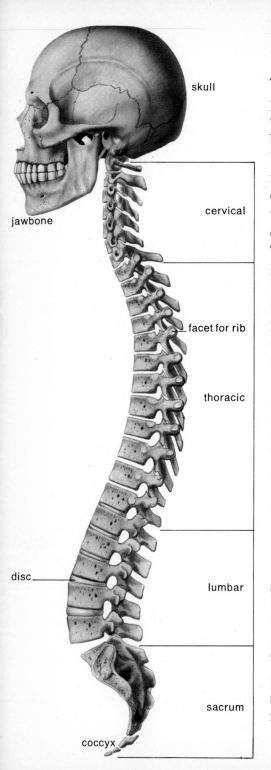

skull

cervical

jawbone

facet for rib

thoracic

disc

lumbar

sacrum

coccyx

The skull is a box of bone that protects the brain, the eyes and the ears. It is made up of various bones all fused together, and is hinged to the lower jaw by a movable joint. The skull swivels on top of the backbone, or vertebral column, which is made of separate bones (vertebrae). They interlock to form a strong, hollow, flexible column through which the spinal cord runs, thus providing a protective skeleton for the spinal cord and a firm support for the trunk of the body. Each vertebra consists of a short rod of bone, above which is a ring enclosing the spinal cord. The vertebrae also have projections to which muscles are attached. Man has 33 vertebrae, and most mammals have about this number, apart from their tail vertebrae. The vertebrae are separated by a cushioning disc of cartilage. If we slip a disc, the soft central part of the disc bulges sideways and presses on a nerve. Each of the various vertebrae is adapted to a particular job. There are seven neck (cervical) vertebrae. The first one, the atlas, fits into and allows movement with the skull to enable the head to be nodded. The second, the axis, lets the head turn on the backbone. The other cervical vertebrae provide attachment for neck muscles. All seven have a hole on each side, through which a blood vessel passes. The 12 chest (thoracic) vertebrae have transverse processes (containing facets) for connection with the ribs and a long pointed neural spine. The five waist (lumbar) vertebrae are big and strong with large projections for the attachment of back muscles. Thoracic and lumbar vertebrae let us bend forwards, backwards and from side to side. The five sacral vertebrae in the hip region are fused to form a strong attachment for the pelvis. At the end, four fused tail (caudal) vertebrae form the coccyx, which is the vestige of a tail.

Man has 12 pairs of ribs, which are joined at the back to the spinal column and form a cage to protect the lungs and heart. The upper seven pairs are joined at the front to the breastbone (sternum), the next three are attached to the seventh rib, and the last two, called the floating ribs, are unattached. Males and females have the same number of ribs and bones, although some differ in size and position.

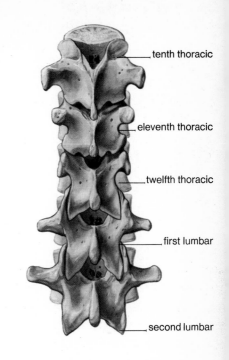

tenth thoracic

eleventh thoracic

twelfth thoracic

first lumbar

second lumbar

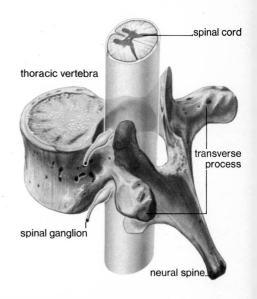

spinal cord

thoracic vertebra

transverse process

spinal ganglion

neural spine

There are three types of muscles in the body. The most common is the striped muscle, also called the voluntary or skeletal muscle, which is composed of specialized cells (muscle fibres) with the ability to contract. Each is controlled by the will for moving the limbs, lifting weights, turning the head and opening the mouth. Striped muscle forms the main flesh of the body and the meat of animals we eat. Many striped muscles are spindle-shaped and are attached at each end to the bones by tendons, which are very strong unstretchable cords of tissue. The calf muscle is attached to the heel by the Achilles tendon which feels almost like a bone, and the fingers are moved by tendons attached to muscles in the arm. These tendons can be seen on the back of the hand if the fingers are moved. In animals tendons are very strong but as they are destroyed by heat, cooked meat easily comes away from the bone.

Striped muscles contract (pull) and then relax. Because they must return to their normal position before they can pull again, these muscles are always in pairs which can pull in opposite directions. They lie across the joints so that their actions cause movement of a

Muscles of the body by 16th & 17th century artists. Those on the right are by Leonardo da Vinci

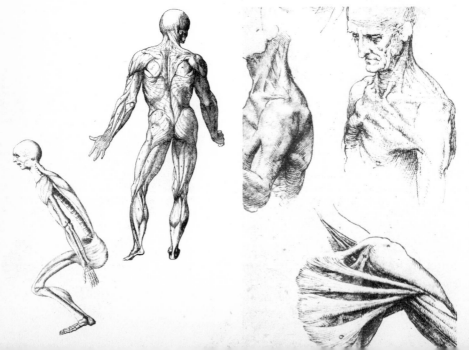

limb. For instance, two muscles are required to bend the arm at the elbow. First there is the biceps, which contracts when the arm is bent and is attached at one end to the radius and at the other end by two tendons to the shoulder bone. The other is the triceps, a smaller muscle which is attached to the elbow at one end and to the shoulder and humerus by three tendons at the other end and contracts to straighten the arm. A striped muscle only contracts when it receives a message to do so, which is sent along motor nerves from the brain.

The second type of muscle is the unstriped, or involuntary, muscle, found in the walls of the intestines, in the bladder and in the blood vessels. It is controlled not by the will but by the autonomic nervous system and functions without our being aware of it: for example, the waves of muscular contractions and relaxations (peristalsis) that force food through the alimentary canal. Only if there is an infection or a blockage do we feel spasms of pain.

The third type of muscle occurs only in the heart. It, too, is not controlled by the will and never tires; the heart goes on beating, without any pauses, for the whole of our life.

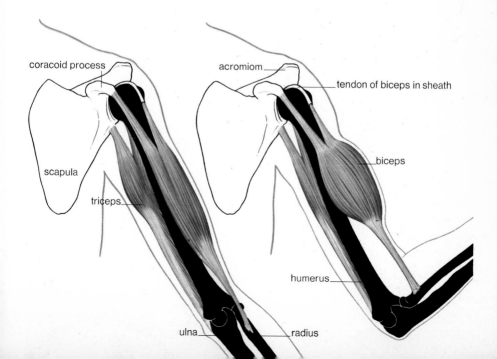

Man's first teeth, called milk or baby teeth, are deciduous – they are shed and replaced by permanent teeth. They are not usually visible at birth but appear when the baby is about seven months old; the full set of 20 takes another two years or so to appear. When a child is about six or seven years old these teeth gradually lose their roots and are pushed out by the permanent teeth growing up from buds present since birth, but it is several years before they are all replaced. There are 32 permanent teeth. The last to come are the wisdom teeth which may not appear until a person is well in his twenties, but sometimes do not grow at all.

The teeth of fishes, amphibians and reptiles are continually re-placed by fresh ones as they wear out. The teeth of plant-eating mammals keep growing to replace wear at the top, but this does not happen in other mammals. The exposed part of a tooth (crown) is covered with a layer of white enamel, the hardest tissue in the body. The long root is embedded in a bony socket in the jaw and is covered with a layer of cement. Beneath the enamel and the cement is the hard bony dentine, which makes up the main bulk of the tooth. At the

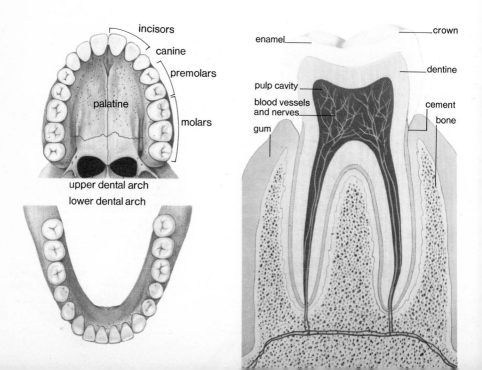

incisors
canine
premolars
palatine
molars
upper dental arch
lower dental arch

enamel
crown
dentine
pulp cavity
blood vessels and nerves
cement
bone
gum

centre is the soft pulp containing blood vessels and nerves which enter through small holes at the tips of the roots.

Unlike some mammals, man has the same number of teeth in each jaw. At the front of the jaw there are two pairs of incisors, wedge-shaped cutting teeth which rodents use for gnawing. Next to these on each side of the jaw is one canine (dog or eye tooth) for tearing food. In flesh-eating animals this is a long curved tooth used for stabbing or holding the prey. Next to the canines are the premolars and molars for grinding and chewing. These cheek teeth have several raised parts (cusps) and, usually, two roots. The first two, the pre-molars, have two cusps and the three molars behind them have four cusps. The molars are not represented in the milk teeth.

The teeth of man are more liable to decay than those of other mammals especially in civilized countries where the diet contains a large quantity of sugar. A layer called dental plaque, which contains mucus, calcium salts and bacteria, tends to form on the surface of the teeth. If it is not removed by regular brushing, decay begins as a result of the acids produced by the bacteria.

The English artist, Thomas Rowlandson, is noted for his comic view of life in the 18th century

A FRENCH DENTIST SHEWING A SPECIMEN OF HIS ARTIFICIAL TEETH AND FALSE PALATES.

Although man can survive for several days without food or water, he cannot live more than a few minutes without air. Air is taken into the body during breathing, and is one of the important functions of respiration. Respiration is a term covering all the processes that help to free energy from our food so that we can make use of it. This is done by using oxygen from the air to break down sugars to release energy; carbon dioxide and water are produced as waste materials.

Respiration goes on in every living cell, so oxygen has to be taken to the cells, and carbon dioxide taken from the cells to the air. To do this, oxygen is extracted from the air by means of the lungs and then carried to all the cells of the body by the blood system. Waste carbon dioxide gas is collected and expelled from the body through the lungs.

Passing air into and out of the lungs is called breathing. During breathing the chest expands and air is sucked into the lungs. In fact, the chest and lungs work like bellows, and breathing is controlled by the movement of the rib cage. When breathing in, muscles arranged across the ribs contract and pull the ribs upwards and outwards, and at the same time muscles of the diaphragm contract and flatten out the rib-cage floor. Both movements enlarge the space the lungs can occupy and so suck in the air. When breathing out all these muscles relax, so the ribs fall back into place and the diaphragm flattens. The space the lungs occupy is made smaller and air is pushed out again. This air contains the carbon dioxide that has been released from the blood.

Air reaches the lungs by way of the nose and mouth, where it is filtered and warmed, down the wind-pipe (trachea) and bronchial tubes and into smaller branches of these tubes, which end in the air sacs (alveoli) of the lungs. Oxygen passes into the blood through the extremely thin walls of the alveoli. Carbon dioxide passes from the blood into the alveoli and is removed from the body when we breathe out. There are up to 700 million alveoli in the lungs and if they could be opened out, their surface would nearly cover a tennis court.

Of the air breathed in, only about one-fifth is oxygen and of this about a quarter is used. The nitrogen, which makes up about four-fifths of the air, is breathed in and out again unchanged.

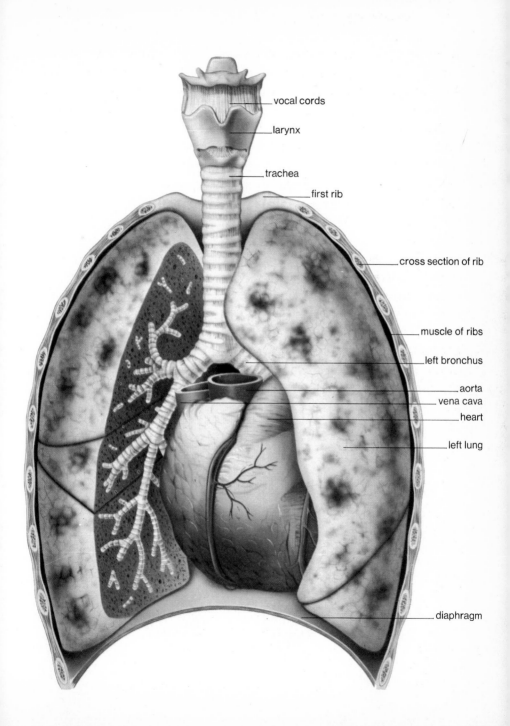

vocal cords

larynx

trachea

first rib

cross section of rib

muscle of ribs

left bronchus

aorta

vena cava

heart

left lung

diaphragm

When a person breathes gently, about 0·5 litre of air is taken in with each breath, but deep breathing can take in up to 2·5 litres. Everyone breathes about 10 000 litres in and out every day. We breathe the least during sleep and the most during intense activity such as running, because there is more carbon dioxide in the blood when the body is active and it is necessary to breathe more deeply to remove it. This build-up of carbon dioxide makes us breathe again if we try to hold our breath. Even in deep breathing the lungs are only partly emptied, and they never collapse completely unless badly damaged.

The body can deal with too much carbon dioxide, but too much oxygen can act as a poison, causing 'oxygen drunkenness'. On the other hand, lack of oxygen may cause a blackout with very little warning. This sometimes happened to pilots in early aeroplanes when they flew too high, because they did not carry oxygen equipment.

The lungs are very delicate and must be well protected. Large particles in the air are caught in hairs in the nostrils, and smaller ones in a sticky fluid (mucus) produced in the bronchial tubes. The mucus would itself soon block the tubes if it were not removed by cilia, tiny

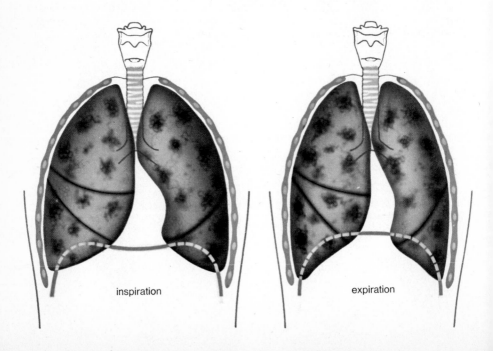

inspiration   expiration

hair-like structures attached to the lining of the tubes. The cilia move back and forth in such a way that the mucus is pushed to the back of the throat and swallowed. When a person has a cold too much mucus is produced, which has to be coughed up or sneezed out. Any irritation or obstruction in the bronchial tubes stimulates coughing to try to remove the trouble. Tobacco smoke may irritate the breathing passages and cause coughing; it may also paralyze the cilia. The epiglottis, a flap at the back of the throat, prevents swallowed food passing down the bronchial tubes. If it does not close, food 'goes down the wrong way' and we start coughing to clear the obstruction. A hiccup occurs when the diaphragm muscles contract suddenly. This cannot be controlled. Nor can a sneeze or cough always be held back, but it is possible to partially control breathing. This is necessary in order to speak. Air is expelled from the lungs and used to vibrate two vocal cords in the voice box (larynx or Adam's apple). Each person has a different speaking voice because the shape of the passages in the nose and mouth also affect the final sound produced, and make an individual 'voice-print'.

Different kinds of breathing. Left: *a sneeze.* Below: *diver with aqualung.* Right: *opera singers*

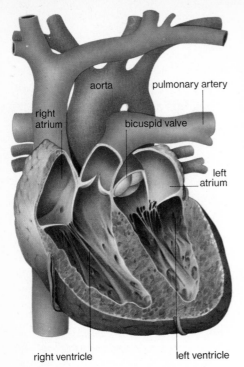

aorta

pulmonary artery

right atrium

bicuspid valve

left atrium

right ventricle

left ventricle

The heart is the pump which does the vital job of circulating the blood throughout the body. The tubes (blood vessels) which carry blood from the heart are the arteries. The blood vessels which return blood to the heart are the veins. A man's heart is about the size of his fist. It is made of powerful muscle which continually contracts and relaxes to produce the heartbeat. The heart is really two pumps side by side. Each pump sucks blood from veins into a collecting chamber, the atrium, or auricle, which then pushes the blood into the ventricle below it. The ventricle pumps the blood under high pressure into arteries. Both ventricles pump at the same time which causes the pulse.

The heart rate varies considerably. In an adult man it is about 70 beats per minute, but in a baby it may be nearly twice as fast. The smaller an animal, the faster its heart beats – about 500 per minute in a small bird and about 20 in elephants. Heartbeat is controlled by a pace-maker nerve which receives nervous impulses from the brain. Vigorous exercise, strong emotions and fear temporarily increase the beat of the heart.

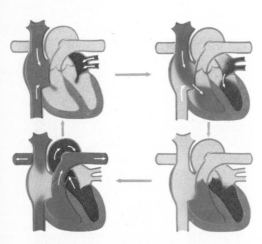

oxygenated blood (red) enters the left atrium and leaves from the left ventricle to the aorta
deoxygenated blood (blue) enters the right atrium and leaves from the right ventricle to the pulmonary artery.

The right ventricle pumps blood that has given up its oxygen to the tissues (deoxygenated blood) through arteries to the lungs. Here it collects oxygen (becomes oxygenated) and returns through veins to the left ventricle to be pumped to the rest of the body before returning to the right side of the heart again. This double circulation is necessary because oxygenated blood leaves the lungs at too low a pressure to pass around the body fast enough to supply the tissues with as much oxygen and food as they need.

Between the auricles and ventricles and at the exits of the heart are small flaps of skin acting as one-way valves to prevent the blood going the wrong way. Any defect in these valves can seriously affect the health. Another serious defect is a hole-in-the-heart, which is a gap between the right and left sides of the heart, usually between the auricles. It allows oxygenated and deoxygenated blood to mingle, so the tissues never get as much oxygen as they need. The heart itself needs a supply of oxygen, which it receives from the coronary artery. A blockage of this vessel can cause a heart attack.

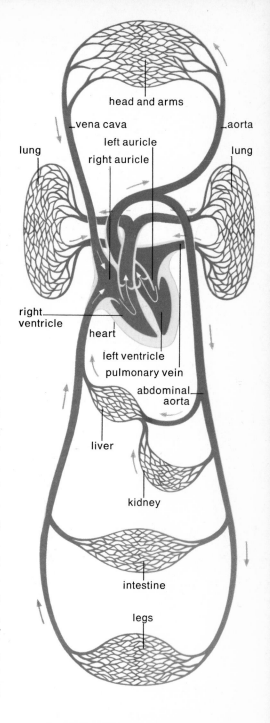

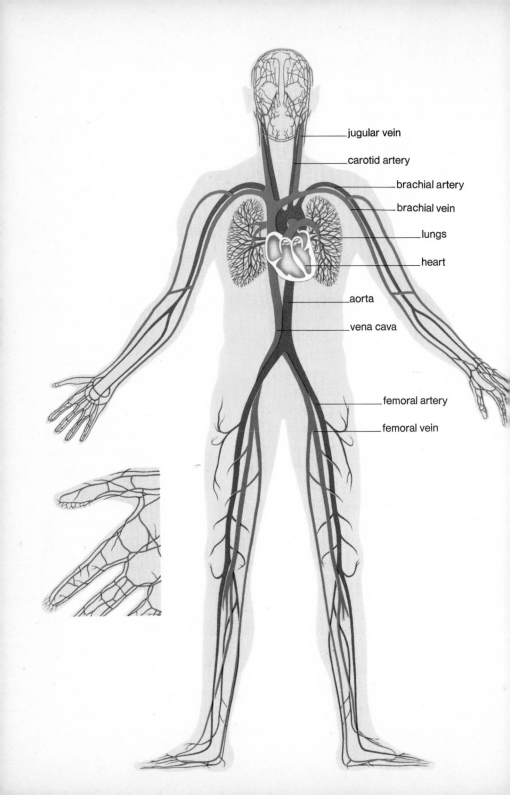

jugular vein

carotid artery

brachial artery

brachial vein

lungs

heart

aorta

vena cava

femoral artery

femoral vein

Blood travels around the body by means of a network of blood vessels that extends throughout the body. It delivers food and oxygen where needed, and carries away waste products. It also helps to distribute heat so that the body is kept at a uniform temperature, fights off infection, heals wounds, and carries chemicals called hormones, which affect growth and other important processes. Blood leaves the heart through large arteries, then travels through smaller and smaller ones and finally reaches networks of minute blood vessels, the capillaries, which have a diameter of only a few thousandths of a millimetre. Capillaries have extremely thin walls through which a two-way exchange takes place: food and chemical substances pass into the body cells and waste products pass from them into the bloodstream.

On its return journey to the heart the blood travels through veins, and as it has lost much of its oxygen, it appears bluish as it flows through them. At different stages of its journey the blood collects necessary materials and unloads waste products. For instance, in the lungs carbon dioxide passes out of the blood and oxygen passes into it, in the small intestine glucose is picked up and carried to the liver, and waste proteins are transported by the blood from the liver to the kidneys. The carotid arteries carry blood from the heart to the head and neck and the jugular veins return deoxygenated blood to the heart and lungs.

Until about three hundred years ago doctors thought that blood flowed through the body with a kind of ebb and flow, like tides. In the 1620s an English doctor, William Harvey, discovered that blood circulates around the body through the system of arteries and veins. The blood travels in a closed system, that is, it always remains within the blood vessels. Arteries are thick-walled tubes into which the blood flows under high pressure from the heart. If an artery is cut, a lot of blood will be lost. Most arteries are deep in the body, but some are nearer the surface and it is in these places that a pulse can be felt. Veins have thinner walls as they usually carry the blood under low pressure after leaving the capillaries and have one-way valves which help keep the blood flowing towards the heart.

An adult man has about 6·2 litres (11 pints) of blood in his body. Blood is not a simple liquid, but consists of a fluid component called plasma in which are floating cells (corpuscles), red cells (erythrocytes), white cells (leucocytes) and small granules called platelets. There are many more red cells than white. Plasma is a pale straw-coloured liquid consisting of water in which are dissolved or suspended a variety of compounds including proteins, dissolved food such as glucose and amino acids, hormones, waste materials such as urea, and salts such as common salt which gives blood its slightly salty taste.

When a drop of blood is examined under a microscope a large number of tiny discs can be seen. These are the red blood cells, which are thinner at the centre and fatter around the rim. Each cell measures about seven micrometres across and about two micrometres at the rim. (A micrometre is one thousandth of a millimetre.) There are about 4 500 000 red cells per cubic millimetre (about 200 or 300 000 000 in one drop of blood); an adult man has about 25 000 000 000 000 red cells. These cells are red because they contain a red pigment – a protein called haemoglobin which contains iron.

Left: *16th century man thought blood flowed like this.* Right: *red blood cell (front and side views)*

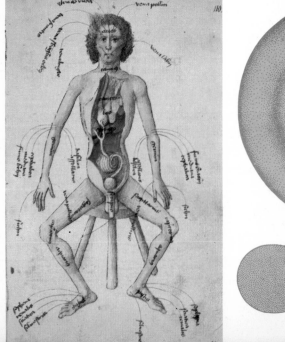

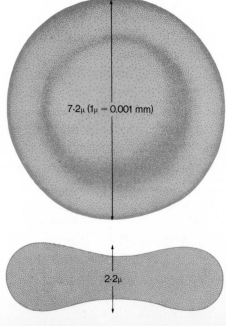

7·2μ (1μ = 0.001 mm)

2·2μ

Haemoglobin combines readily with oxygen to form a compound called oxyhaemoglobin in parts of the body where there is a lot of oxygen (in the lungs), and just as readily loses oxygen where it is scarce (in the body tissues). Most carbon dioxide travels dissolved in the plasma. This exchange process enables the body cells to continue to carry on an active life.

The red cells of man are unusual in that they lose their nucleus while they are developing. This probably explains why they live for only 100 to 120 days. Dead ones are broken down and most of the material lost, but the iron is stored in the liver until needed. New cells are formed in the bone marrow. Blood itself does not reach every cell of the body. Food and oxygen are delivered to the cells by a colourless liquid called lymph, which consists of plasma and white cells that squeeze out of the capillaries and bathe all cells and tissues. Lymph also forms the fluid in blisters. It protects and feeds the body and gradually drains back into the veins along narrow lymph ducts which in some places, mainly the intestines, armpit and groin, connect with the lymph nodes (glands) where white cells are produced.

Human blood photographed under a microscope. Dark areas are nuclei of white blood cells

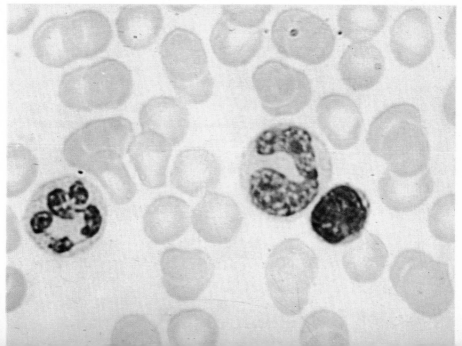

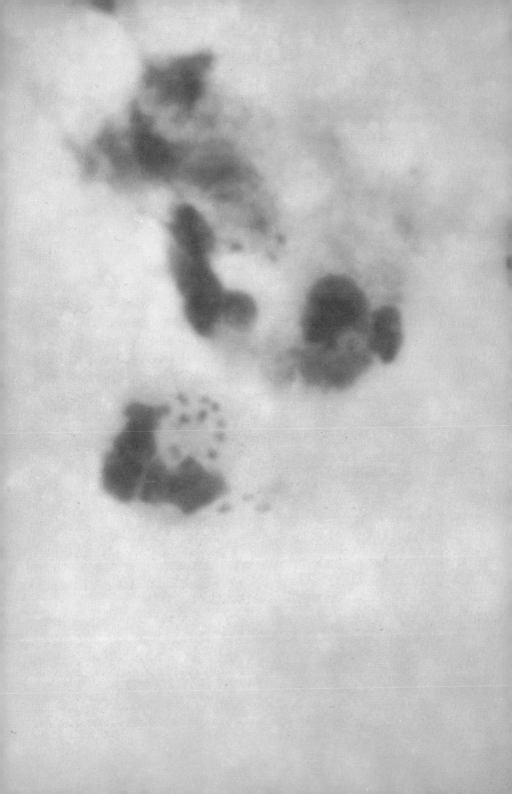

The white blood cells (leucocytes) provide the body's defence against disease. They are not really white, but colourless. They are often called white corpuscles, but unlike red corpuscles they have a nucleus and are capable of independent motion.

White cells vary in size but most of them are about four times as big as red corpuscles and there are fewer of them, one to every six or seven hundred red cells. Although they are mostly carried in the bloodstream in the same way as the red corpuscles, when they touch the wall of a blood capillary they can squeeze between the cells making up the thin walls of a capillary and enter the lymph system. They move in the same way as an amoeba, by pushing out parts of themselves called pseudopodia. As they travel in the bloodstream they are constantly ready to attack any foreign matter, especially dead cells or harmful bacteria. They dispose of these by flowing around them and enclosing them, in the same way as an amoeba engulfs its food. White cells have a short life of only a few days, because the bacteria they consume poisons them. For example, an infected finger may become swollen with pus. Pus is a yellowish fluid consisting of dead leucocytes which have been poisoned by the bacteria that caused the infection.

There are several kinds of white blood cells. One type fights off disease by producing chemical substances called antibodies which kill the disease-producing bacteria and neutralize the bacterial poisons (toxins). Antibodies can be artificially produced in the bloodstream by vaccination or inoculation with dead disease organisms or neutralized toxins. This is what happens when someone is immunized against diseases such as diphtheria, smallpox and polio. White cells, like red corpuscles, must be constantly renewed to keep up a steady number: about 700 to the cubic millimetre. They are made in the red bone marrow, the lymph glands and in the spleen.

Another vital part of the defence system are the platelets, minute round or oval grains, much smaller than the cells and about a tenth as numerous, which play an important part in making the blood clot.

White blood cells, magnified 1600 times, enclose foreign particles within their protoplasm. This is an important defence mechanism of the body for destroying harmful invading bacteria

When a person cuts himself the blood soon sets into a jelly, and later hardens into a scab. The blood platelets, plasma and damaged tissue combine to form a network of minute threads which traps the corpuscles and forms a clot. But there is a rare disease called haemophilia in which the blood is unable to clot. Haemophilia is a hereditary disease, which means people are born with it and can pass it on to their children. It usually affects only males but it can be passed on by females. It is well known because Queen Victoria was a carrier and many members of her family were haemophiliacs.

In the 19th century doctors thought that people who had lost blood from an injury could be helped by being given a transfusion of blood from another person. The result was almost always disastrous. As soon as the donor's blood went into the patient's bloodstream the patient's own blood reacted against the foreign blood and clotting occurred. Eventually doctors discovered that people had different, genetically-determined blood groups, only some of which would mix with others. The four main groups are called A, B, AB and O to indicate whether certain substances called antigens are present in the

Queen Victoria and her family in 1880. Some of them, and their descendants, were haemophiliacs

blood. For instance, group AB has antigens A and B but group O has neither. If group A blood is given to someone without antigen A (that is, group B or O), that person's blood will clot the transfused blood. Group O people can give their blood to anyone, but can only receive it from other group O people, whereas group AB people can receive blood from anyone, but only give it to other AB people. Another well-known blood group is the rhesus factor, so-called because it was first found in rhesus monkeys. A high proportion of humans are rhesus positive; they have the rhesus antigen. Although blood falls into the four main groups no two persons (except identical twins) have blood which is absolutely identical, and in an emergency group O blood is used for transfusion.

The body as a whole, and especially the white blood cells, reacts strongly in the same way as the antigens of the different blood groups, to expel any foreign matter which gets into the body. This protects against bacteria and viruses, but makes the transplantation of organs such as heart or kidney very difficult, because without very careful matching the patient's body will reject the transplanted organ.

The females of the British royal family who carried haemophilia, and the males who died of it

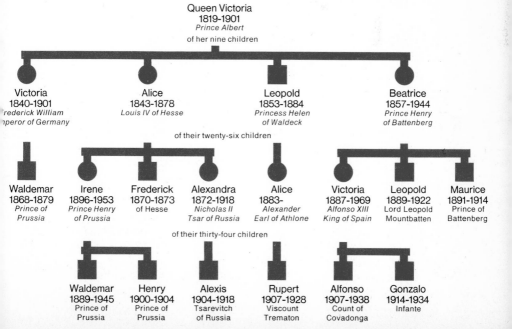

Food is the raw material used by living organisms for making new body tissue and, after combining with oxygen, for providing energy. The energy content of food is the amount of heat – measured in Calories or joules – produced when the food combines with oxygen. The different types of foods are classified chemically as proteins, fats and carbohydrates; also necessary for healthy growth and development are vitamins and minerals. Two-thirds of the body is composed of water; this is continually being lost, and is replaced by the fluids we drink and also by the water in many foods.

Protein molecules are complex chemical compounds containing nitrogen, carbon, hydrogen and oxygen. They are all built up of standard units called amino acids. Protein makes up part of every cell, and is also the substance of muscles and very important chemicals, called enzymes, that are found in all cells. Animal protein from milk, meat and cheese is made up of amino acids, all of which are useful to us.

Fats and carbohydrates contain carbon, hydrogen, and oxygen. Fats are greasy compounds made of units called fatty acids and

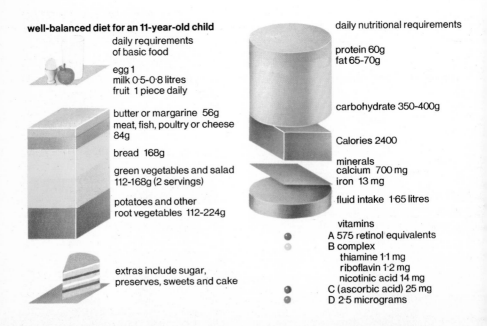

well-balanced diet for an 11-year-old child

daily requirements of basic food

egg 1
milk 0·5-0·8 litres
fruit 1 piece daily

butter or margarine 56g
meat, fish, poultry or cheese 84g

bread 168g

green vegetables and salad 112-168g (2 servings)

potatoes and other root vegetables 112-224g

extras include sugar, preserves, sweets and cake

daily nutritional requirements

protein 60g
fat 65-70g

carbohydrate 350-400g

Calories 2400

minerals
calcium 700 mg
iron 13 mg

fluid intake 1·65 litres

vitamins
A 575 retinol equivalents
B complex
  thiamine 1·1 mg
  riboflavin 1·2 mg
  nicotinic acid 14 mg
C (ascorbic acid) 25 mg
D 2·5 micrograms

glycerol. Fat produces more energy per unit weight than carbohydrates. It is the white substance around muscle, and we usually eat it with meat, but dairy foods such as milk, butter and cheese also contain fat, and so does margarine. Cooking oils are liquid fats. Fats are stored under the skin to prevent heat loss. Extra carbohydrates can be converted to fat in the body, which is why too much carbohydrate makes us gain weight. Carbohydrates are sugars and starches. Food that tastes sweet contains sugar, and starches are found in bread and potatoes. These are fuel foods, which are important because they yield energy when they are burned in respiration.

The body also needs certain other chemicals. Besides sodium and chlorine in salt, small amounts of other elements are absorbed in food, including potassium, calcium, magnesium, iron and traces of zinc, copper and iodine, all of which are essential for good health. According to where they live, or their income, people have very different diets. Many North Americans eat a great deal of meat, the Japanese get much of their protein from fish and the Chinese and Indians eat mainly rice and vegetables.

It is not possible to stay strong and healthy by only eating meat, bread, sugar and fat. Many years ago sailors on long vovages were fed these foods and often became ill with a disease called scurvy, which led to loss of weight, bleeding and sometimes death. It is now known that scurvy is caused by a deficiency of vitamin C, which is found in fruits and vegetables. But long before this was discovered, good captains knew that a ration of lemon or lime juice kept their crews healthy.

Early in this century it was found that foods contain minute quantities of certain substances vital to life and health, most of which the body cannot make for itself. These are called vitamins. Some are very complicated chemical compounds, but most can now be made artificially. So far, about 50 types of vitamin have been identified, many of which are essential in a healthy diet. Vitamin A is most important for clear vision and general well-being. Liver is rich in vitamin A, but it is found also in fish, meat, milk, butter and some fruits and vegetables. The body can make vitamin A from the orange pigment in carrots and similar vegetables. The vitamin B complex includes 12 different chemicals, found in eggs, milk and its products, wholemeal flour and vegetables. A shortage of these vitamins affects the whole body, including the skin, nervous system and blood. Vitamin C, as well as preventing scurvy, helps wounds to heal and some scientists believe that large quantities may prevent colds. Fruits and uncooked vegetables are rich in vitamin C, but if they are old or overcooked most of the vitamins are lost. Vitamin D, essential for the growth of bones and teeth, is found in fish liver oil and milk. Unlike the other vitamins, this is one that the body can make for itself, if there is plenty of sunlight. A deficiency of both sunlight and a good diet leads to rickets, which causes bone softening and deformity, a disease from which people in sunny lands seldom suffer.

Essential as they are to good health, a lifetime's supply of all needed vitamins would weigh only a quarter of a kilogram. Because vitamins can be manufactured they are often advertised as additions to our food, but a normal well-balanced diet provides all the necessary vitamins.

the principal vitamins: sources and deficiency effects

common name	chemical name	occurrence	effect of deficiency
A	retinol	cod liver oil, liver, eggs, milk, butter, green vegetables, carrots	night blindness
B₁	thiamine, aneurine	yeast, wheat germ, stoneground bread, lean meat, rice bran	beri-beri which affects nervous system and heart
B₂	riboflavin	yeast, liver, kidneys, cheese, meat	disorders of eyes, skin and mucous membranes
B₆	pyridoxine	yeast, wheat germ, liver, potatoes	anaemia, dermatitis, retarded growth
part of B complex	nicotinic acid	yeast, wheat, bread, liver, meat, fish, chicker mushrooms, rice bran	pellagra which affects skin, mucous membranes and nervous system
(H)	biotin	yeast, peanuts, chocolate, peas, mushrooms	uncertain
part of B complex	pantothenic acid	yeast, liver, beans, mushrooms, peanuts	uncertain
part of B complex	folic acid	yeast, liver, milk, green vegetables	anaemia
B₁₂	cyanocobalamin	liver, meat, milk	pernicious anaemia
C	ascorbic acid	fresh fruit and vegetables, especially lemons, oranges, blackcurrants, tomatoes, rose hips	scurvy which affects hair, teeth, gums, skin, joints, and wound healing
D	calciferol	cod liver oil, eggs, milk, butter, sunshine on skin	rickets, leading to bone deformities
E	tocopherol	wheat germ, green vegetables, vegetable oil	uncertain
K	phylloquinone	green vegetables, liver, tomatoes	spontaneous bleeding

Food is useful to the body only when it has passed into the blood-stream. To do this it must be broken down into molecules small enough to pass through the walls of the intestine and dissolve in the blood. This conversion of large molecules of food to small ones is called digestion. It takes up to 24 hours for food to pass along the 10-metre long digestive tract (alimentary canal). Digestion is brought about by enzymes which act as catalysts – substances which speed up chemical reactions without being used up themselves. Enzymes are present in fluids produced in the upper half of the alimentary canal, and they convert starch and complex sugars to glucose, proteins to amino acids, and fats to fatty acids and glycerol.

Food is first ground up (masticated) and mixed by the teeth and tongue and moistened by saliva. Saliva contains an enzyme which begins to digest starch into sugars. The swallowing of food is the only digestive process partly under voluntary control; the rest is carried out automatically by the regular squeezing action of muscles in the alimentary canal. The food is lubricated by mucus which is produced by the walls of the canal and also helps to stop the canal from digesting itself. The food quickly travels down the gullet (oesophagus) to the stomach where it is churned up. The stomach, with a capacity of over a litre, produces gastric juice which contains enzymes and dilute hydrochloric acid. The acid kills most of the bacteria in the food and provides the best medium for the stomach enzymes to begin digesting protein. Because solid foods remain in the stomach longer than liquids, milk is curdled so that it will stay long enough to be digested. The partly digested food passes next into the duodenum, the first part of the small intestine – a coiled tube about eight metres long and as wide as a man's thumb. Digestion is completed by the action of bile from the liver and enzymes from the pancreas and small intestine itself to form a mixture of simple sugars, amino acids, fatty acids and glycerol. The amino acids and simple sugars are absorbed by the blood in the walls of the second part of the small intestine (ileum), but most of the fatty acids and glycerol re-combine to form fat which passes into the lymph. The remaining undigested food, in the form of a liquid mass, passes to the large intestine, or colon.

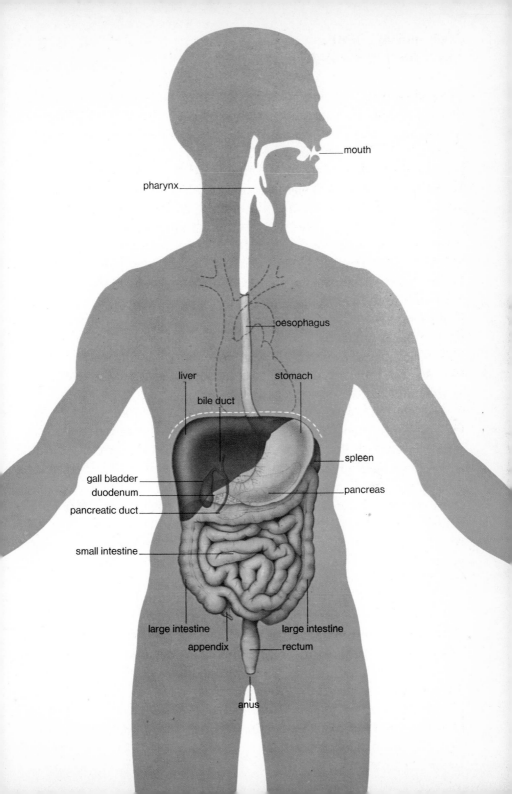

mouth

pharynx

oesophagus

liver

stomach

bile duct

spleen

gall bladder

pancreas

duodenum

pancreatic duct

small intestine

large intestine

large intestine

appendix

rectum

anus

The small intestine widens into the large intestine (colon), where most of the water from undigested food is absorbed. From there the remaining material, now semi-solid, passes into the rectum where it collects before being ejected from the anus as faeces. The caecum forms the junction between the small and large intestines.

The inner surfaces of the intestines are folded, which increases the surface for absorption, but the small intestine has, in addition, millions of tiny finger-like villi containing numerous blood capillaries, which give an even greater area for absorption. The blood that has absorbed food from the intestine passes to the liver before returning to the heart, because one of the main functions of the liver is to regulate the foods in the blood.

The liver, the largest organ of the body, is just below the diaphragm next to the stomach and consists of several deep red lobes. Each lobe is a loosely-packed mass of cells between which the blood percolates. Because the jobs it does need a great deal of energy, the liver is one of the main heat-producers of the body.

The liver converts any remaining sugars into glucose, which provides the body with energy. Any glucose in the blood that is not immediately required by the body is converted to a starchy insoluble substance called glycogen, most of which is stored in the liver. The liver also reduces fatty acids and glycerol to a form that can provide more energy for the body. Some of the breakdown products of amino acids from the digestion of protein are poisonous, so these have to be converted by the liver to the less harmful substance, urea. Urea is removed from the blood by the kidneys, and is then passed out in urine.

Another important function of the liver is in the manufacture of bile. The liver removes iron from dead red blood cells and the rest of the cell material then becomes the part of the bile called the bile pigments. Bile passes down the bile duct into the small intestine where another part, called bile salts, acts like a detergent by breaking up the large droplets of fat so that they can be digested more easily. Bile is a bitter-tasting green-yellow liquid, and is stored until needed in the gall bladder. The liver also makes the chemicals that help to clot blood.

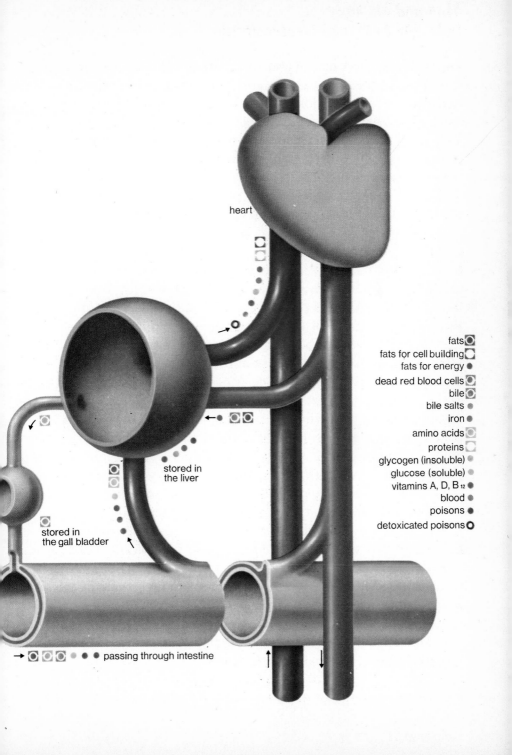

heart

fats ◉
fats for cell building ◎
fats for energy ●
dead red blood cells ◉
bile ◉
bile salts ●
iron ●
amino acids ◉
proteins ◎
glycogen (insoluble) ●
glucose (soluble) ●
vitamins A, D, B₁₂ ●
blood ●
poisons ●
detoxicated poisons ◯

stored in
the liver

stored in
the gall bladder

→ ◉ ◉ ◉ ● ● ● passing through intestine

The pancreas, between 10 and 15 cm long, is shaped like a carrot and is pinkish-grey in colour. It is in the abdomen behind the lower part of the stomach, and is joined by a pipe, or duct, to the first part of the small intestine (duodenum), into which it pours pancreatic juice. This juice is a mixture of enzymes that act on carbohydrates, fats and proteins, helping to break them down into the basic food units that can be absorbed into the blood through the walls of the intestines.

Another vital product of the pancreas is insulin, the hormone which controls the amount of the glucose (a sugar) in the blood and its use by the tissues. Insulin is produced in a part of the pancreas called the islets of Langerhans. The normal level of insulin in the blood is 0·1% but people with diabetes do not produce enough insulin to reach this level. Thus an excessive amount of sugar is present in the blood which cannot be used by the tissues. In this condition much of the sugar passes out of the body in the urine. Diabetes can be controlled by injections of insulin extracted from cattle and other animals, and also by carefully limiting the amount of sugar in the diet. Diabetes affects about one person in a hundred in the United States, for example, and there are probably many more who have the disease without knowing it. It occurs mainly in countries where people are well fed. The islets of Langerhans in the pancreas also produce a hormone called glucagon, which controls the conversion of glucose into glycogen (the chief storage carbohydrate of animals).

The spleen, a spongy mass about the size of the heart, lies near the stomach not far from the pancreas, but it does not belong to the digestive system. The spleen in a foetus manufactures red blood cells, but in adults it seems to act as a reservoir for holding and releasing blood and it is thought to play some part in regulating blood pressure. It filters off used red blood cells and it may make some of the white blood cells. It may also have other functions but they are not yet identified. In some illnesses the spleen becomes enlarged. It is not a vital organ and if it is removed by surgery other parts of the body, such as the liver, will take over its functions.

The discovery of insulin revolutionized the treatment of diabetes. Two young Canadian doctors, F. G. Banting and C. H. Best, took only eight months to develop the drug at Toronto in 1922

The chemical processes which take place in the body produce waste material, and the kidneys are responsible for filtering off much of it. The kidneys also control the amount of water, minerals and other substances in the blood. These substances should be at a certain level and not vary according to how much we eat or drink. All the extra material is removed with the waste through the bladder as urine.

The kidneys look like a pair of very large red bean seeds attached to the back of the body cavity just below the ribs. They are usually embedded in fat which helps to protect them from heavy blows. Blood reaches the kidneys under high pressure from the heart via the dorsal aorta and renal artery, and passes into tiny capillaries in the kidneys. Here much of the water and dissolved substances are forced out of the capillaries and into about a million microscopic blind-ended tubes in each kidney. As this fluid passes down the tubes the useful materials and nearly all of the water pass back into capillaries surrounding the tubes. The rest of the water and the harmful material is the urine and it passes down a thin tube (ureter) to the bladder where it is stored. The wall of the bladder stretches so that it can hold

Left: *section of a kidney*. Right: *enlarged section showing renal tubules*

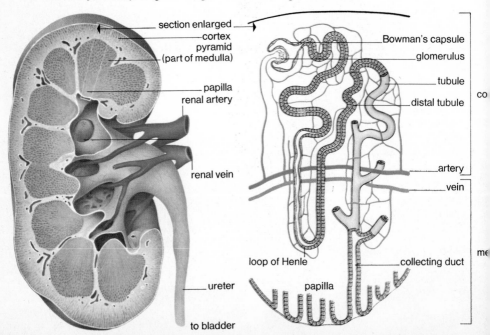

section enlarged
cortex
pyramid
(part of medulla)
papilla
renal artery
renal vein
ureter
to bladder

Bowman's capsule
glomerulus
tubule
distal tubule
artery
vein
loop of Henle
papilla
collecting duct
co
me

about a litre of fluid; once the limit is reached the wall contracts and the urine is expelled through the urethra. Medical researchers have developed machines which can take over when the kidneys are seriously damaged or fail to function properly.

The main content of urine is urea, a waste substance produced in the liver from unwanted amino acids, but drugs and other harmful substances are also filtered out of the blood in the kidneys and appear in the urine. That is why a urine test reveals how much alcohol a person has been drinking, as well as other indications of a person's health. It can reveal diabetes, if there is sugar present, or pregnancy when particular hormones appear in urine.

An adult produces about 1·5 litres of urine per day, but the amount varies widely. For instance, because the kidneys control water in the blood, far more urine is produced when a person has been drinking a lot, or in cold weather when no water is lost by sweating. Some desert animals are very good at saving water, many of them getting all they need from their food alone. Birds and insects do not produce liquid urine. Instead of urea in solution, they produce solid uric acid.

A special machine can take over the work of kidneys which do not function properly

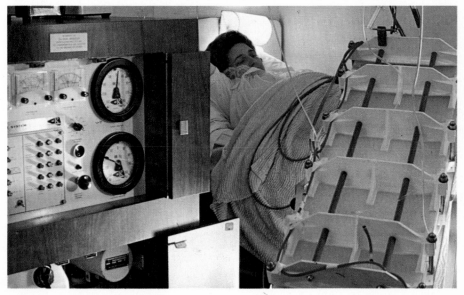

The skin has three main jobs: it is a waterproof protector that prevents excessive loss of moisture from the body and wards off infection, it helps to maintain a regular body temperature of around 37°C (98·4°F) and it is a sensitive detector of touch as well as of heat and cold. Skin is about one mm thick and is composed of two layers are tougher and drier than those underneath and the uppermost thicker dermis. The epidermis contains many layers of cells. The top layers are tougher and drier than those underneath and the uppermost are dead cells that flake off in very small pieces. Dandruff is a mass of such cells that flake off from the scalp. The dead cells contain keratin, the material found in hair and nails. As the top layers wear away new ones grow up from the bottom to replace them. On the soles of the feet the dead layer is thick and horny to withstand wear.

The epidermis contains hair, sebaceous glands and sweat glands, all of which thrust deep down into the dermis. Each hair grows up from the bottom of a tiny pit called a follicle. Although it grows from this living root, the part above the skin – the hair we see – is a series of dead horny cells. A hair grows for about four years, then it falls out

Section cut through skin to show a hair root

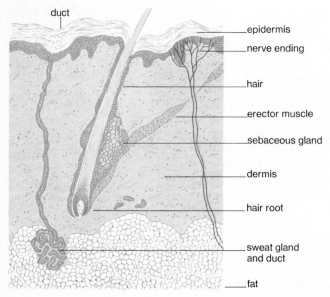

duct

epidermis

nerve ending

hair

erector muscle

sebaceous gland

dermis

hair root

sweat gland
and duct

fat

and a new one replaces it. Hair colour is determined by pigments produced within the hair. When it ceases to produce pigments hair turns grey. Attached to each hair follicle is a tiny muscle which makes the hair stand on end. This is obvious in animals when they are alarmed or angry but they also raise their hair to keep themselves warm. The sebaceous glands secrete a fatty substance that helps to keep the hairs and the surface of the skin in good condition. Sweat glands are very important as their secretion, sweat, helps to keep the skin cool.

Nails are hard, protective, horny outgrowths from the epidermis, like the hoofs and claws of animals.

The dermis contains a great number of small blood vessels, muscles, nerves and fat cells. It is soft but tough and elastic and its surface is furrowed with deep ridges. On the surface of our hands and feet, and especially at our fingertips, these ridges are not filled in by the epidermis and it is this that makes the whorls and loops of fingerprints. When the skin is burned, blisters form. These are pockets of clear watery fluid, lymph, which form between the dermis and epidermis. The lymph helps to protect the skin against the effects of the burn.

An area of skin, from a woman's calf, magnified 60 times

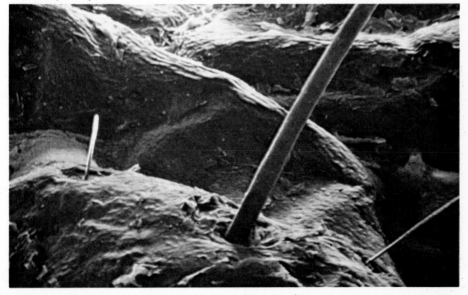

Skin colour is due to a brown pigment, melanin, in the epidermis,
which protects the body from the
harmful effect of the sun's ultra-
violet rays. The different amounts
of melanin cause the shades of
colour of various races. When
people tan it is because they produce more melanin in response to
sunshine. Freckles are small scattered patches of melanin. A few
individuals have no melanin:
they are albinos, with completely
white skin and hair, and eyes that
look bloodshot because there is
no pigment in the iris to mask the
red blood vessels.

Insects, crustaceans and reptiles are cold-blooded, which
means their body temperature
varies with the temperature of
their environment. Flies, for instance, are active when it is hot
but drowse when it is cold. Birds
and mammals are warm-blooded
with a constant temperature at
which body processes work best,
so they are just as lively in cold as
in hot weather. In man this temperature is about 37°C (98·4°F).
There is a thermostat in the brain
which regulates temperature by
sending messages throughout the
body. The skin with its capillaries,

An African mother and her albino child

sweat glands and hairs plays an important part in maintaining a steady temperature. The new process of thermovision translates temperature differences in the skin into different colours recorded on a temperature map. This method enables doctors to detect some kinds of skin cancer. The evaporation of sweat from the skin causes much heat loss from the body, and people sweat most in the hottest weather in order to keep cool. Sweat consists mostly of water, with small amounts of salts, and a man working in the tropics can lose a litre of water an hour by sweating. Excessive loss of water from the body can cause heatstroke. When one is hot the skin capillaries are at their widest, so that they can carry more blood to the skin. This heat loss by radiation makes the skin look red. In very cold weather the capillaries shrink to prevent heat loss so that the flow of blood almost stops and the skin becomes white, or even bluish if the blood darkens due to lack of oxygen. In animals the hairs stand on end to trap a layer of air next to the skin; gooseflesh in man is a similar attempt to trap a layer of air by raising the hairs on the skin.

Thermovision, a new method of making temperature maps, or thermographs, can help scientists detect heat variations in human skin

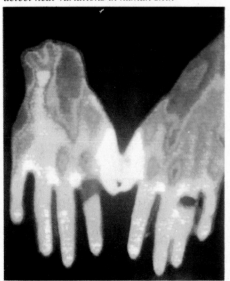
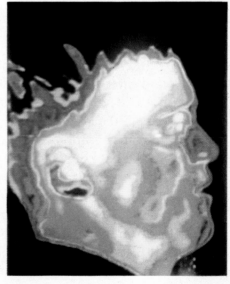

Even the simplest living things must adjust to their surroundings to avoid danger, find their food and recognize their own kind. In man this is the function of the nervous system. With the help of the hormones it ensures that all our body systems work together and that our sense organs gather information about what is happening all around us so that we can make the right kind of adjustments.

The central part of the nervous system is the brain and the spinal cord. The spinal cord is about one cm wide where it leaves the base of the brain, and gradually tapers as it passes down the body through the backbone. Pairs of nerves, from both the brain and spinal cord, run to all parts of the body. Twelve pairs travel from the brain, mostly going to the head and throat, and 30 more go from the spinal cord to the trunk, arms and legs. The brain and spinal cord have a small fluid-filled space at their centres, but the nerves themselves are solid.

The nervous system consists of two parts. The voluntary nervous system lets us send out orders through motor nerves to muscles to carry out actions such as standing, walking, running and talking, which have to be learned after birth. It also controls messages received through sensory nerves from the skin and sense organs such as the eyes, ears and taste buds of the tongue. Millions of sense organs in the skin are connected to the voluntary nervous system: they are sensitive to pain, heat and cold. The autonomic nervous system keeps our body organs working smoothly without our knowing or thinking about it. It automatically controls the heartbeat, the activities of the stomach and intestines and production from glands of liquids such as sweat and saliva. Part of this system is controlled by a pair of nerves from the base of the brain, while the rest is supplied by branches from other nerves which combine to form two parallel cords running along each side of the backbone.

Although the autonomic and voluntary systems are separate there are links between them, but it is not always clear which is operating. Most of the functions controlled by the autonomic system cannot, of course, be affected by orders from the brain, but some individuals, for example those practising yoga, seem to be able to exert some control over their rate of breathing, and even over their rate of heartbeat.

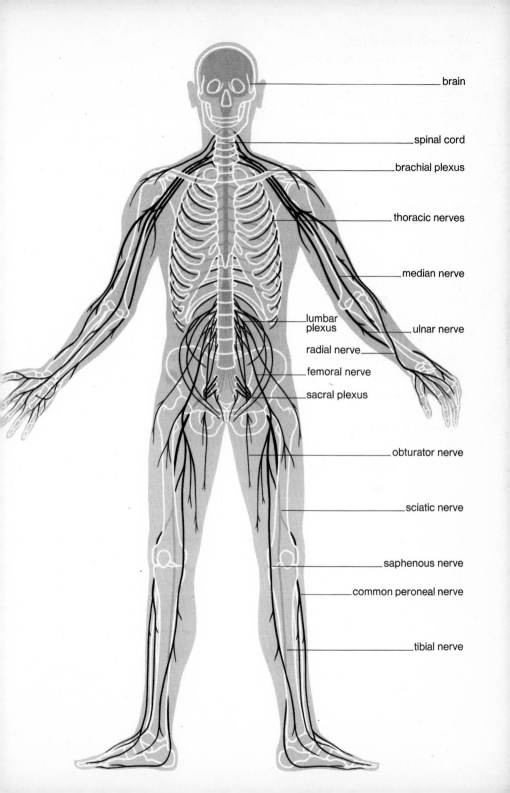

brain

spinal cord

brachial plexus

thoracic nerves

median nerve

lumbar plexus

ulnar nerve

radial nerve

femoral nerve

sacral plexus

obturator nerve

sciatic nerve

saphenous nerve

common peroneal nerve

tibial nerve

A nerve may look like a strand of whitish cotton or thickish cord or it may be too thin to be easily seen. It is made of a number of nerve fibres, each of which forms part of a nerve cell. The whole nerve is covered by an insulating sheet. A nerve cell consists of at least one long fibre and a cell body containing the cell nucleus. Sensory cells pick up and pass on messages from the sense organs through receptors; connector cells link the different parts of the central system, and motor cells carry orders to the muscles. The message is sent along the nerve fibre in the form of a nervous impulse and then passed from one nerve cell to another through very fine branches which interlock with, but do not quite touch, those of the adjoining cell. The tiny gap between two nerve cells is called a synapse, and the impulse travels across the synapse by the release of a chemical substance at the nerve endings. The impulse is carried along the fibre as a small electric current. The cell bodies are found only in, or just outside, the brain and spinal cord, mainly in the grey matter in the centre of the cord and in the outer part of the brain. This means that nerve cells are

Photomicrographs of a nerve cell body (left) *and a motor nerve ending* (right). *Compare with the diagram on the opposite page*

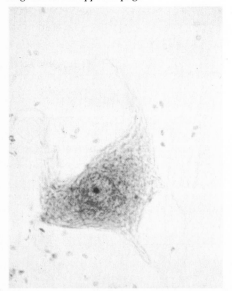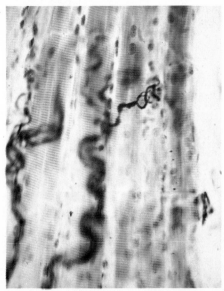

much larger than most other cells – a single fibre may run right up from a toe to the backbone.

If your hand touches something uncomfortably hot, an impulse in the form of an electric current passes from the sense cell in the hand along nerve fibres to cell bodies just outside the spinal cord. The impulse is carried into the grey matter of the spinal cord, then transferred to connector nerve fibres which take it to the brain. The brain may decide that the hand is in danger and an impulse will travel along other connector cells to motor cells in the spinal cord grey matter. Fibres from these cells carry the impulse to arm muscle cells which contract to move the hand away. If the danger is very great, the message 'short circuits' in the spinal cord, and the hand is withdrawn before the brain can deal with the problem. This is called a reflex action. Most actions are more complicated. Even deciding to write one's name needs a vast number of sensory and motor impulses. These processes are very rapid. It takes less than one-fiftieth of a second for an itch impulse to travel from toe to brain.

Direction of impulses in a sensory nerve cell

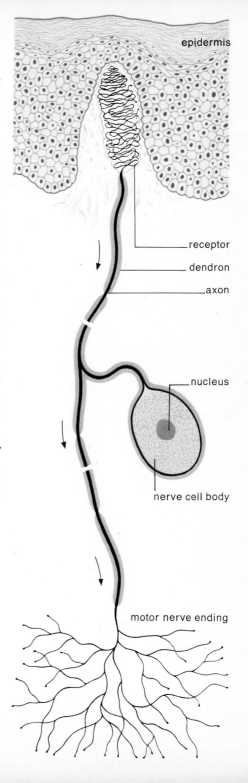

epidermis

receptor

dendron

axon

nucleus

nerve cell body

motor nerve ending

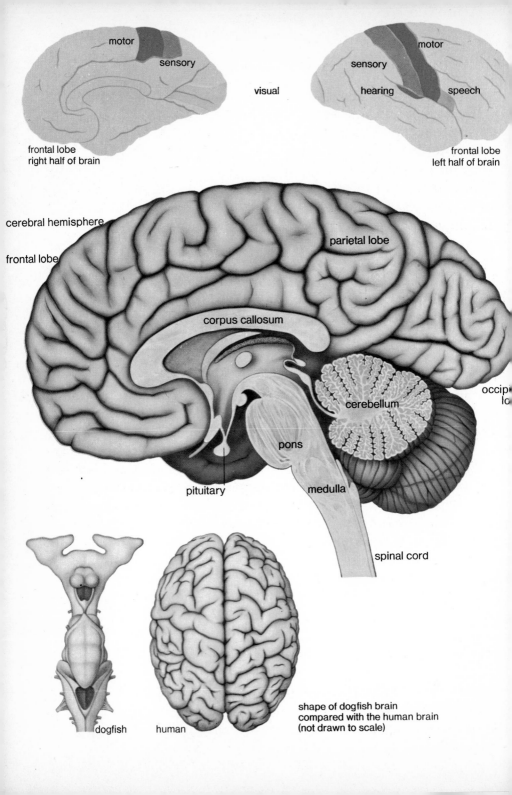

motor

sensory

visual

motor

sensory

hearing

speech

frontal lobe
right half of brain

frontal lobe
left half of brain

cerebral hemisphere

frontal lobe

parietal lobe

corpus callosum

occip
lo

cerebellum

pons

pituitary

medulla

spinal cord

dogfish

human

shape of dogfish brain
compared with the human brain
(not drawn to scale)

The brain acts as a control for all the various activities of the body. These include not only conscious activities, such as whether to stand or sit, or laugh or cry, but involuntary activities.

Man's brain has three main parts. At the top of the backbone where the nerves bunch together to enter the skull is the medulla. This controls the unconscious activities of our bodies, which are under the influence of the autonomic nervous system, and include breathing, heartbeat and digestion. Above the medulla is the cerebellum which controls our sense of balance and assists the coordination of muscular activities. These controls have to be learned, but they are so well learned that they become automatic. Above the cerebellum is the largest part of the brain, the cerebrum, which is divided into two halves (hemispheres) and makes up 70% of the whole brain and nervous system. This part of the brain is far more developed in man than in any other animal. Its surface is a complicated pattern of wrinkles giving it a very large area into which are packed an enormous number of nerve cells, forming an outer layer called the grey matter. This outer layer of the cerebrum is associated with the capacity for intelligent behaviour, speech, learning, imagination, memory and decision-making. The grey matter also receives and interprets impulses concerned with conscious activities. Beneath this layer is a thicker layer, the white matter, consisting of nerve fibres. The right half of the cerebrum controls the activities and movements on the left side of the body, and vice versa. There are special areas of the brain that control hearing, smell and sight. A small but very important part of the brain is the pituitary body at the base of the brain in the medulla. It produces important hormones that control many of the other hormone systems of the body, and is involved in growth, development and reproduction.

The brain of man is larger, in comparison with his body size, than that of any other animal. It grows rapidly at first and then more slowly. If brain cells die or are damaged they are not replaced by new ones as are the cells of other organs. The brain is suspended in liquid and well protected by its bony box, the cranium, but any damage to it can seriously affect a person's personality and abilities.

In some ways an eye is like a complicated camera. A camera has a lens which focuses the light onto a sensitive film in a dark box. The amount of light entering is adjusted by altering the hole (aperture) through which it enters. The 'box' of the eye is the more or less spherical coat, or sclera, part of which is the white of the eye. This tough layer is protected by a bony socket to which it is attached by three pairs of muscles which move the eye around. The eye is further protected by fat, by the eyelids which wipe the front clean, and by tears which are secreted from ducts behind the top eyelid to wash away dust particles and kill bacteria. Tears are produced all the time, but they usually drain away so we do not notice them. When the eye is irritated more tears are produced and the excess drains down a channel into the nose. When we cry tears are produced at a faster rate than they can drain away.

Inside the sclerotic layer is a dark layer called the choroid, which stops blurring when light is reflected within the eye, and also contains blood vessels that supply the inner part of the eye. Beneath the choroid is the retina, the light-sensitive layer of the eye, which is

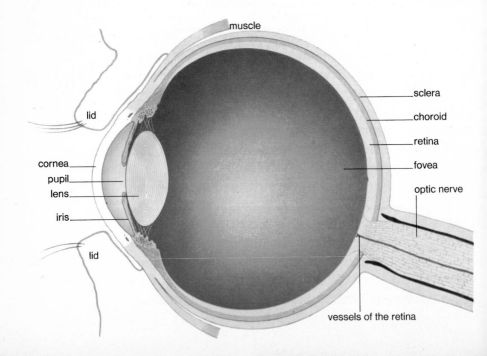

several cells deep. It contains two types of light-sensitive cells: the rods, which give black-and-white vision and are sensitive even in poor light, and the cones which give colour vision and function best in bright light. The most sensitive spot on the retina is the yellow spot (fovea) which contains many cones.

The transparent bulging front of the eye is the cornea, and it helps to bend the light rays so that they fall onto the retina. Fine focusing is carried out by the lens. A camera is focused by moving the lens backwards or forwards, but in the eye the lens focuses the light by varying its shape, by means of the contraction and relaxation of the muscles to which it is attached. In front of the lens is the iris, which we can see as the coloured part of the eye. It controls the size of the central hole (pupil) through which the light passes (the weaker the light the wider the pupil). The black colour of the pupil is due to the choroid, which can be seen through the pupil. The inner part of the eye is filled with a clear liquid, which is rather watery in front of the lens but more jelly-like behind it.

Left: a small section of the retina, greatly magnified. Right: rods and cones in the retina are linked to nerve cells, which connect nerve fibres to the optic nerve

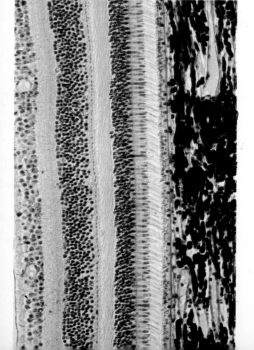

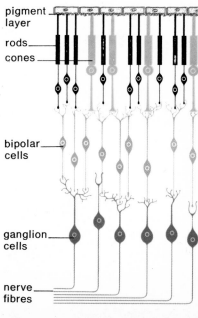

pigment layer

rods

cones

bipolar cells

ganglion cells

nerve fibres

Man and other primates have two eyes set at the front of the head
which each give a slightly different view of the same object. These are
turned into a three-dimensional (stereoscopic) view, to judge distances
accurately. Stereoscopic vision is most developed in hunting and tree-
dwelling primates. Because man's ancestors were probably tree-
dwellers who had to land accurately when jumping, this may be why
he has such good stereoscopic vision. If one eye is closed it is much
harder to judge distances. In hoofed animals, such as deer and
antelope, the eyes are set on each side of the head so that they cover a
wider field but give a much less detailed view. Man and monkeys,
among the mammals, are able to see different colours. Some of the
cones in the retina are sensitive to red light, some to green, others to
blue. A number of men, but only a very few women, are at least partly
colour blind, which means they cannot distinguish one colour from
another. The most common form of colour blindness is the inability
to tell red from green.

Cones are more sensitive than rods to colour and to minute differences in detail. At one point at the back of the retina, immediately

Red-green colour-blind people cannot see the object in the centre of the dotted circle

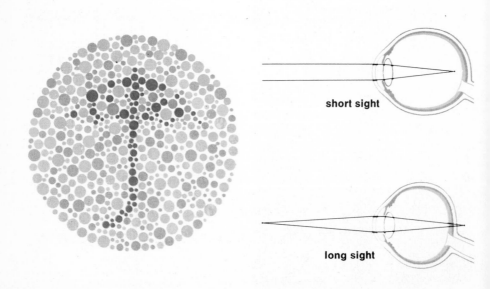

short sight

long sight

opposite the lens, is the yellow spot (fovea), an area that contains only cones and is not covered by blood vessels and nerve fibres. This is the area of clearest vision, but it means that in poor light, when the cones do not operate, a better view of an object may be obtained by looking slightly to one side of it so that the light rays fall on the rods. It is because the cones do not operate in poor light that we see mainly in shades of black and white under these conditions.

When vision is imperfect it may be because of long sight, when nearby objects are focused behind the retina, or short sight, when distant objects are focused in front of the retina. Both conditions can be corrected by artificial lenses, which were first invented in Europe in the 13th century. Long sight needs convex lenses which bend the light towards the centre of the cornea. Short sight needs concave lenses which bend the light away. As people age, their lenses and the surrounding muscles gradually lose their elasticity, so they cannot adjust to different distances. This condition is called presbyopia and can be corrected by bifocal lenses. The lower part of these lenses is used for reading, the upper part for looking at distant objects.

How a garden appears to the short-sighted. A concave lens (circle) makes distant blooms clear

Our ears help us to perceive and locate the direction of a sound, but practice is necessary. This is why babies take some time to learn to do it. An ear consists of three distinct parts: the outer, the middle and the inner ear.

The outer ear is the flap (pinna) we can see, which acts as a sound-collecting trumpet, and a short twisting passage which leads to a stretched membrane, the tympanic membrane or ear-drum, inside the ear.

The middle ear is an air-filled chamber with an opening into the back of the nose and throat through the three-cm long Eustachian tube. If the pressure outside the ear-drum changes rapidly, as in going up or down in an aeroplane, or even a lift, the pressure is equalized by air entering or leaving through the Eustachian tube which opens when we yawn or swallow. When catarrh partly blocks the tube, the effect of changing pressure is particularly annoying.

Inside the middle ear, which is about five cm long, are three tiny bones: the hammer (malleus), the anvil (incus) and the stirrup (stapes), arranged so that when the ear-drum vibrates it rocks the

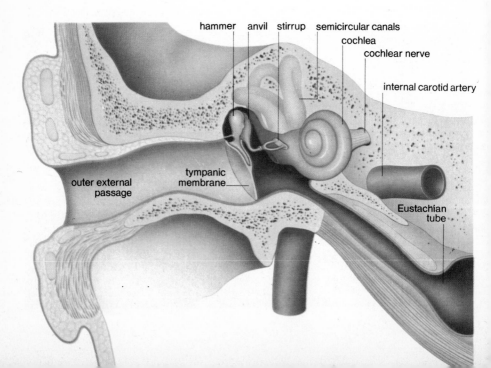

hammer, which rocks the anvil, which rocks the stirrup which passes the vibration on through an oval membrane (window) to the inner ear. The movement of the comparatively large drum is multiplied about 20 times before it reaches the small oval window.

The inner ear has no air in it and is even deeper inside the skull than the middle ear. It has two distinct parts with quite different jobs to do.

The cochlea, the hearing organ, is an anti-clockwise spiral which looks like a snail shell. It consists of channels which are filled with a fluid and contain the membranous organ of Corti. Vibrations from the oval window are transmitted through the fluid and press on the organ of Corti. This stimulates sensitive hair cells, which pass impulses to the brain. It is the job of the brain to translate and interpret these impulses into the various sounds we hear. The human ear can detect a wide range of sounds, from the low-pitched note of thunder to the high-pitched whistle of a kettle.

The second part of the inner ear consists of three semicircular canals, concerned with balance, which resemble three spirit levels arranged in three different planes at right angles to each other.

section through the cochlea

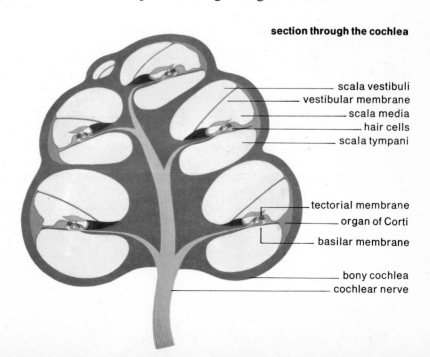

scala vestibuli
vestibular membrane
scala media
hair cells
scala tympani

tectorial membrane
organ of Corti
basilar membrane

bony cochlea
cochlear nerve

Vibrations are carried through the bones of the skull and the bones of the middle ear. Sounds carried through the skull produce a slightly different effect from those going through the ear. That is why a person receiving a tap on the head hears a loud noise and a tape recording of your voice sounds odd. Since hearing depends on cells that respond to the pressure of vibrations and have developed from the touch cells of the skin, it is possible to 'hear' different vibrations from tuning forks held against the hairs on the forearm.

A person with good hearing is sensitive to a sound frequency range from about 20 to 20 000 hertz (cycles per second). For comparison, middle C on a piano has a frequency of 256 hertz and top C 4096 hertz; a healthy hearing range is therefore 10 to 11 octaves. This is well below a rat's hearing which goes up to 90 000 hertz.

As people age, their hearing range gets narrower. People over 60 years old, for example, find it more difficult to hear birds singing. This sort of deafness is due to the nerve fibres withering but, although it is a nuisance, it does not usually deteriorate into complete deafness. A battery hearing aid is a microphone that picks up sound and con-

A wide variety of mechanical aids are used to help deaf children learn to speak

verts it into electrical energy, which is then amplified and changed back into sound by transistors and a tiny loudspeaker. The sound is conducted either along a tube to the ear-drum or through the bones of the skull by a vibrator. Not so long ago, children who were born deaf were deaf-mutes and grew up without learning to speak. However it is now possible to teach deaf children to talk.

The semicircular canals in the inner ear help us to balance. They are set at right angles to each other, so that no matter how the head is moved – up, down, left or right – the fluid in the canals will also move. The pressure of this fluid stimulates the sensitive hairs within the canals to send messages to the brain. If a person spins rapidly and then stops, the fluid will continue moving, causing giddiness. This can be prevented by suddenly jerking the head in the opposite direction.

Two structures, the utriculus and sacculus, join the cochlea to the canals and act as gravity detectors. They contain grains of chalk which move about as we tilt, and trigger sensitive hairs. Continuous up-and-down movement over-stimulates the brain which is what causes sea sickness.

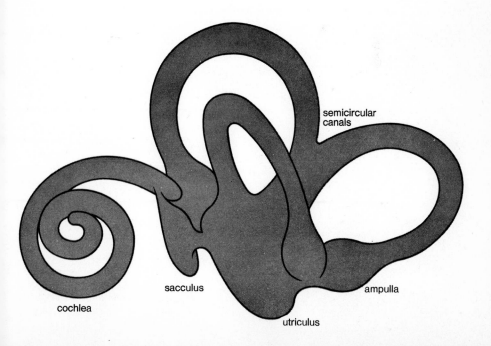

semicircular canals

sacculus

ampulla

cochlea

utriculus

"Ah! Bisto"

Man and his body
155   taste, touch, smell

Smell and taste are perceived by senses which respond to certain chemical stimuli, and a large variety of substances can be detected by these senses. Taste buds are found in special areas on the tongue. When food touches the taste buds, they send nerve impulses to the brain, which are identified as different tastes. Man can distinguish four tastes: bitter at the back of the tongue, sour at the sides and salt and sweet at the front. Taste is only possible if the substance can dissolve in water, although only a small amount is required. Taste is often at least partly due to smell; some tastes are lost if the nose is blocked by catarrh. Smells are detected by two small groups of sensory hairs occupying a small area at the top of the nose chamber. Molecules of a strong-smelling substance dissolve in the mucus in the nose chamber and stimulate the hairs, which send messages to the brain by special nerves. Humans have a poor sense of smell compared with most mammals. Dogs have a keen sense of smell and recognize friend, enemy and territory as much by smell as by sight. Every-

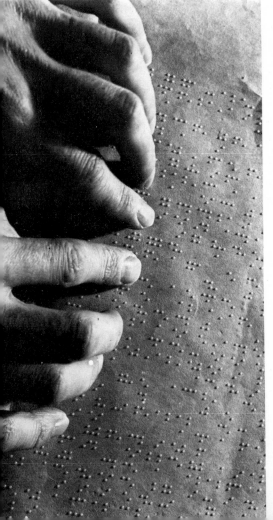

Braille, raised letter printing, enables the blind to read through touch cells in the fingers

one, except identical twins, has a different smell, and a bloodhound can follow a scent with ease. There are many different smells but it is thought that these are only different combinations of four basic types of smell: fragrant, burned, acid and rancid. Even unlikely substances which cannot be eaten, such as iron and rubber, have distinct tastes and smells. It is thought that the shapes of the molecules in matter determine their taste or smell.

Sight, hearing, balance, smell and taste are all called the special senses because they are located in particular groups of sensory cells. The sense of touch is a combination of five sensations: contact, pressure, heat, cold and pain. The receptor cells sensitive to touch are present all over the skin, although the number of particular sensory cells varies from place to place. For instance, there are far more touch cells on the lips and fingertips than on the back of the hand. The sensory cells for heat and cold usually show whether a person is becoming hotter or colder. If one hand is soaked in hot water and the other in cold water, then both are placed in luke-warm water, the water feels warmer to the hand that has been in cold water.

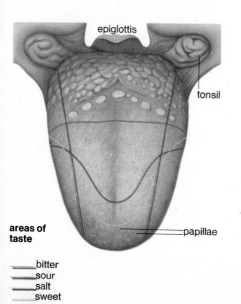

epiglottis

tonsil

areas of
taste

_____bitter
_____sour
_____salt
_____sweet

papillae

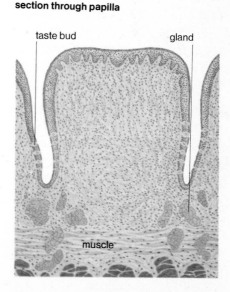

section through papilla

taste bud

gland

muscle

The nervous system is not our only messenger system; there are also chemical messengers in the body called hormones. These substances are made by various glands in different parts of the body and are released straight into the bloodstream. They circulate around the body until they reach the part where they cause a particular process to occur. A well-known example is insulin, produced by the pancreas, which removes excess sugar from the blood.

The most complicated hormone-producing gland is the pituitary, which lies in a little cavity in the skull at the base of the brain. It regulates many other glands, including the thyroid and adrenal glands, and makes several hormones of its own. This is why it is called the master gland. One of the pituitary hormones controls the growth rate of our bones and body tissues. Another controls regulation of the water content of the blood carried out by the kidneys. When a baby is born, the mother's pituitary sends out a hormone that makes the womb muscles contract so that the birth takes place.

The thyroid gland situated in the neck on either side of the windpipe produces the hormone thyroxine, which regulates the rate at which oxygen is used, and also helps control growth. The thyroxine in the blood must be at the right level as an excess causes restlessness and excitability, and too little makes people fat and sluggish.

Next to the thyroid are the parathyroids, four tiny, glands which regulate the amount of calcium and phosphorus in the blood.

The adrenal glands, just above the kidneys, produce various hormones including adrenaline and the corticosteroids. If we are tense, angry or afraid, adrenaline is secreted into the blood and causes increases in the heartbeat, breathing rate and blood pressure. This has the effect of helping us to fight or run away, depending on the circumstances, or even to do better in examinations. The corticosteroids are divided into two main groups: one regulates the amount of carbohydrate in the blood and the other controls the level of minerals in the blood.

The sex organs (the testes in males and the ovaries in females) secrete hormones that stimulate the development of masculine or feminine characteristics, such as beards in men and breasts in women.

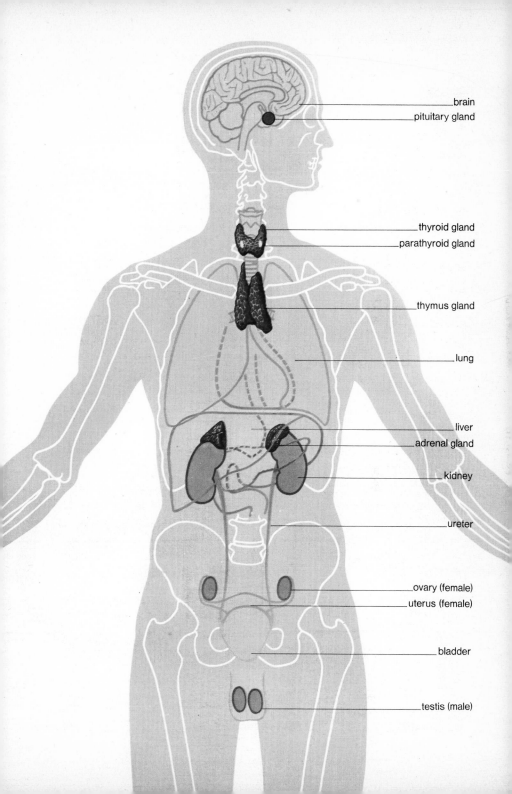

brain
pituitary gland

thyroid gland
parathyroid gland

thymus gland

lung

liver
adrenal gland

kidney

ureter

ovary (female)
uterus (female)

bladder

testis (male)

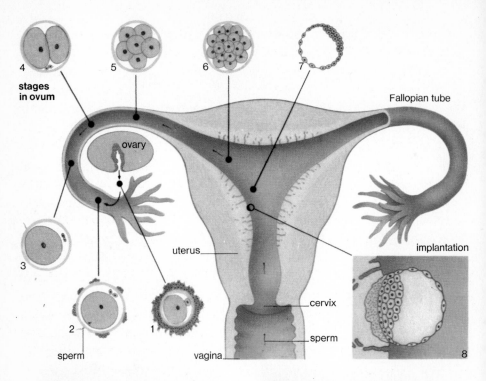

4   5   6   7

stages in ovum

Fallopian tube

ovary

3

2

1

sperm

uterus

vagina

cervix

sperm

implantation

8

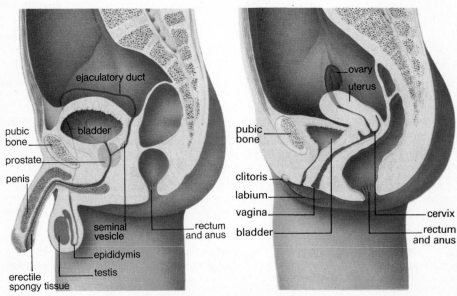

ejaculatory duct

pubic bone

bladder

prostate

penis

seminal vesicle

rectum and anus

epididymis

testis

erectile spongy tissue

ovary

uterus

pubic bone

clitoris

labium

vagina

bladder

cervix

rectum and anus

The offspring of most animals are conceived when a sperm cell unites with an egg cell. Another kind of fertilization takes place in plants after pollination, although plants, and some simple animals, can also reproduce in other ways. The female fish lays thousands of eggs and then the male fertilizes them by spraying them with sperm. Female birds form eggs within their bodies and if they are fertilized by sperm from the male within the body of the female, the eggs develop into chicks. Domestic hens sometimes lay eggs which have not been fertilized. In all mammals the young develop inside the mother's body and are born alive, as distinct from developing outside the mother's body in an egg. Mammals are called viviparous (live-bearing).

Women between the ages of about 13 and 45 produce an egg about once every 28 days which passes from the ovaries to the uterus (womb). For 48 hours the egg is able to be fertilized but after that it dies and is carried away during menstruation. Fertilization, or conception, can occur when the male's penis is pushed into the vagina and ejects millions of sperms. The sperms swim up the vagina and uterus and into the fallopian tube, where one of them may meet an egg and fertilize it. Between humans this sexual act can take place at any time although fertilization does not always occur, but for most mammals there are fixed periods when it happens.

In dogs, a bitch has eggs ready to be fertilized about once every six months. At these times, when she is said to be on heat or in season, she secretes certain substances with a special smell that attracts male dogs. There are similar periods for other animals, varying in length from every three to four weeks in mice to once every two years in elephants. Similarly, the time taken for the full development, or gestation, of the embryo inside its mother varies. In white mice it is three weeks, in humans nine months, and in elephants 21 months.

Many mammals produce several offspring at one birth, but humans usually bear a single child. Sometimes, however, there will be two or more babies at the same birth. Identical twins occur when a single fertilized egg divides in two and each half develops into a separate but identical child. Fraternal twins result from the presence of two eggs in the mother's uterus each of which is fertilized by a different sperm.

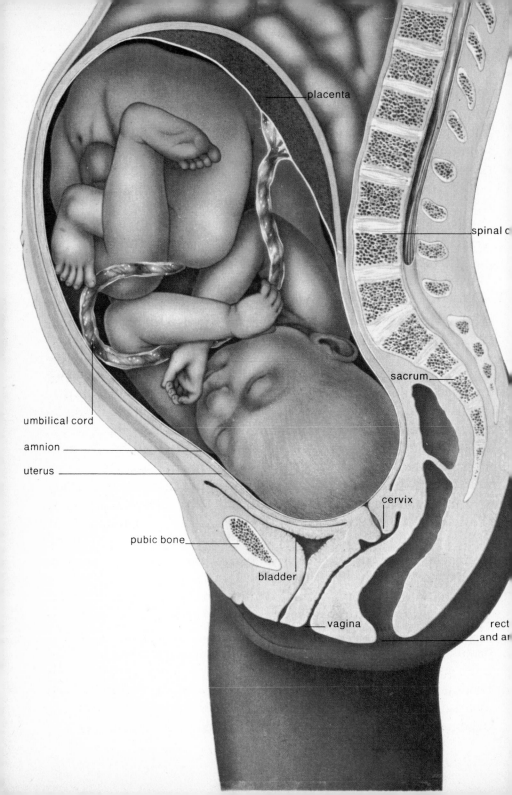

placenta

spinal c

sacrum

umbilical cord

amnion

uterus

cervix

pubic bone

bladder

vagina

rect
and ar

When an egg is fertilized it starts to develop into a new organism. Until it is born it is called an embryo. A bird's egg contains an embryo surrounded by a store of food (yolk) on which the embryo feeds until hatching. The development of a human baby is quite different. It takes place entirely inside the mother.

The fertilized egg starts to divide until it forms a many-celled embryo. This is embedded in the wall of the mother's uterus and until birth it remains attached by the umbilical cord to the placenta, a pad of tissue attached to the wall of the uterus. The embryo's food and oxygen come from its mother's blood through the placenta, and it gets rid of its waste material into the mother's blood in the same way. The embryo is suspended in a fluid-filled sac (amnion) and surrounded by membranes to protect it from physical damage. In its early stages of development the embryo has features that recall more primitive animals. Gradually it takes on a more human appearance and by the eighth week, although still very small, it shows the general features of the baby-to-be. But its head remains large in comparison with the rest of its body. The sex of a baby is determined at conception and is usually not known until its birth. Sometimes doctors can make special tests to find out its sex while it is still in the womb.

After about nine months the baby, now weighing about four kg, is ready to be born. Unlike the baby chick, which breaks out of its shell, the baby cannot make its own way out of the uterus but must rely on the strong muscular contractions of the uterus to push it through the vagina. During these contractions the membranes surrounding the baby are broken. This period, called labour, can vary from less than an hour to several days.

When the baby is born the umbilical cord attaching it to the mother is severed (its end forms the navel) and the baby begins to breathe. The helpless baby depends for many months on its mother for food, which is sucked in the form of milk from her breasts or from a bottle. Not until it is at least a year old can it stand alone and it needs help for many more years before it can take care of itself. Even in primitive societies it is not until a boy is 12 or 13 years old that he is thought fit to hunt and travel alone.

The moment a human egg is fertilized within the mother its sex is decided, and at birth male and female babies are very alike, except for the differences in their sex organs. Each year as they grow the differences become more marked.

In a male the external sex organs are the penis and a pouch of skin beneath it, the scrotum, which contains two oval testes in which the sperms are produced. The testes do not start to function until a boy is in his teens. Before this the penis functions only as a tube for emptying the bladder. After this it also carries the sperm.

The sex organs of a female are inside her body. The entrance to the vagina is surrounded by folds of fleshy tissue. The vagina leads into the uterus which is connected by two tubes to the ovaries, the organs which produce the egg cells. All the eggs that can be produced are present, although very tiny, in the ovaries from birth. After a girl reaches puberty and until she is in her forties she will produce about one egg a month. At the same time the uterus regularly prepares itself for bearing a baby. If the egg is not fertilized the lining of the uterus breaks up and passes out through the vagina with some blood. These regular periods of bleeding are called menstruation. In females the bladder empties by a passage which is quite separate from the sex organs, unlike the arrangement in males in which the excretory and reproductive systems are closely connected.

The sex organs of both boys and girls do not begin to function until the age of 12 or 13. This stage, called puberty, may vary in both individuals and societies, because it is affected by diet and climate. After about the age of 10 the differences between boys and girls become more obvious. In girls the breasts gradually become larger and they go on growing for the next few years. Boys begin to grow hair on the face and their voices deepen. Their shoulders become broader and their hips narrower in comparison with girls. Both girls and boys (the girls usually two or three years earlier) grow hair in their armpits and around their genitals. In both boys and girls the changes at puberty are caused by the secretion of hormones from the pituitary gland. These pituitary hormones stimulate the sex organs to produce special hormones that bring about the physical changes.

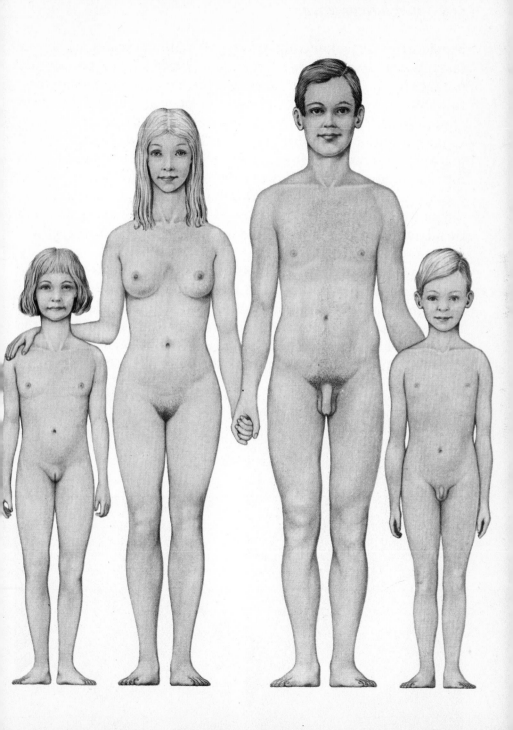

People of every colour, appearance or nationality belong to the same species, *Homo sapiens*. They can all breed with each other and produce fertile offspring. In this sense there is only one human race. In another sense, the different races of man are defined as broad groups of people whose colour or features are so characteristic that it seems natural to group them together. Most experts believe that all mankind is descended from a single original stock and that the differences now existing between the races arose rather late in the evolution of man, through natural selection and adaptation to different climates and surroundings. However, intermarriage, slavery or migration tended to blend the differences so that now even the purest race is mixed.

There have been recent attempts to classify the human population on the basis of the differences in their blood groups. But there are other and more obvious inherited physical characteristics also used to classify races. There are differences not only of skin colour but in the shape of head, nose, mouth, height and kind of hair. Colour alone gives a simple classification into the white (Caucasoid), black

Left: *a Negro of the Diola tribe, Senegal.* Right: *a Nordic girl from Scandinavia*

(Negroid) and yellow (Mongoloid) races.

A more complex classification involves subdivision of these three basic groups. The principal race of Caucasoid stock is the Mediterranean whose characteristics include a long head, narrow nose, thin lips, prominent chin, olive skin and dark hair. Typical Mediterranean types are Spaniards, Italians and Arabs. Other Caucasoids are the Nordics, found in Scandinavia, and the Alpines who live in France, Switzerland and Asia Minor. The races of India are also variants of the Mediterranean type.

The Australoids are a race whose features include wavy hair, darkish skin and flat noses. They are the Aborigines of Australia and the inhabitants of New Guinea and the Pacific islands.

Negroids have long heads, woolly hair, very dark skin and flat noses. The Negroes of western and central Africa are typical examples.

Mongoloids are distinguished by broad heads, straight black hair, yellow or reddish skin and flat noses. Examples are the peoples of central and eastern Asia. The Amerindians, or American Indians, are believed to have descended from an Asiatic Mongoloid stock.

Left: *a Mongoloid boy from Laos.* Top: *a subdivision of the Mediterranean race includes Yugoslavs from Montenegro.* Below: *Australian Aborigine*

Index

Diploid, 15
Diptera, order, 78
Diseases
  from flatworms, 62
  from flies, 78
  from fungi, 20, 21
  from protozoans, 58, 59
  from roundworms, 63
  from viruses, 17
Distillation, 52, 55
Division, in cell, 6, 15, 158
DNA (deoxyribonucleic
  acid), 5, 6, 15, 17, 18
Dodder, 43
Dodo, 98
Dog family, 1, **2**, 117
— rose, 32
— (eye) teeth, 132
Dogfish (rock salmon), 8, 84
Dolphin, 116
Domestic cat, 118
— cattle, 122
— dog, *canis familiaris*, 1, **2**,
  117
Dominant factors (genes), 14
Donkey, 120
Dorsal aorta, 145
Double circulation of blood,
  135
— helix, 5
Douglas fir, **28**, 29
Dracunculus, Guinea worm,
  63
Dragon
  flying, 92
  Komodo, 92
  sea, 87
Dragonfly, 52, 53, 73, **79**
Drone bee, 77
Drosera, sundew, 43
Drugs
  addiction to, 54
  beneficial, 54
  harmful, 54
  from plants, 54
Dry rot, 21
Dryopteris, male fern, 27
Duck-billed platypus, 104,
  105

Ducks, 100, 101
  eider, 99
Duckweed, 31, 37
  great, **37**
Dugong, 116
Dung beetle, 75
Duodenum, 142
Dust bowls, 48

Eagles, 97, 100, 101
Ear
  of bats, 107
  of crocodiles, 94
  of mammals, 104
  of man, 153, 154
Earthworms, 66
Easter lily, 36
Eberthella typhi, bacterium,
  18
Echidna, order, 105, 115
Echinoderms, 80
Ecological communities,
  123, 125
— regions, 123
Ecology, 1, 123
Edentata, order, 115
Edible (Roman) snail, **64**
Eel grass, *Zostera*, 37
Eels, 86, 126
Egg cell, 3, 6, 157
Egg-laying mammals, *see*
  monotremes
Eggs
  of bees, 77
  of birds, 96
  of crocodiles, 94
  of fishes, 85
  of frogs, 90
  of lampreys, 82
  of mammals, 105
  of penguins, 98
  of reptiles, 91
  of toads, 90
Egyptian vulture, 102
Eider duck, 99
Electric ray, 84
Electron microscope, 3
Elephant seal, 116

Elephantiasis, 63
Elephants, 10, 11, 119
Elk, 121
Elm tree, 35
Elodea, pondweed, 37
Elver, 86
Embryo, 8, 158
Emperor penguin, 98, 101
Emu, 98
Endive, 34
Endocrine system, 128
Endoplasmic reticulum, 3, 5
Endosperm, 49
Energy, 39
  chemical, 41
  content of food, 140
  light, 40
  mechanical, 41
  release, 41
Environment, changes in, 7
  pollution of, 127
Enzymes, 5, 39, 140, 142
Eocene period, 11
Eohippus, fossil horse, 11
Ephemeroptera, order, 79
Epidermis
  of leaves, 39
  of skin, 146
  of stems, 42
  of roots, 42
Epiglottis, 134
Epiphyte plants, 44
Epithelial tissue, 128
Equidae, *see* horses
Equisetum, horsetail, 26
Equus, horse, 11
Ericaceae, family, 33
Erythrocytes (red blood
  cells), 137
Euglena, protozoan, 59
Euphrates river valley, 48
European bison, 122
— eel, 86
Eustachian tube, 153
Evaporation
  from plants, 42
  of sweat, 147
Even-toed ungulates, 121
Evergreen plants, 28, 29

Lupin, 32
Lyell, Charles, 10
Lymph, 137, 142, 146
  in blood, 137
  ducts, 137
  nodes (glands), 137, 138
  system, 138
Lynx, 118

Macaque, monkey, 110
Mackerel shark, **126**
Macrocystis, alga, 23
Maggot, 78
Magnoliaceae (magnolia),
  family, 32
Maize, *see* corn
Malaria, 59
Malay fruit bat, 107
Male fern, *Dryopteris*, 27
Malleus, ear bone, 153
Malus, genus, 32
— *pumila*, crab apple, 51
Mammalia, *see* mammals
Mammals, 104
  anteaters, 115
  bats, 107
  carnivores, 117, 118
  elephants, 119
  insectivores, 108
  lagomorphs, 114
  marine, 116
  marsupials, 104, 106
  monotremes, 105
  primates, 110, 111
  rodents, 112, 113
  ungulates
    even-toed, 121, 122
    odd-toed, 120
  see also man
Mammary glands, 104
Mammillaria, cactus, 33
Mammoth, woolly, 10, 11,
  119
Man
  and apes compared, 111
  evolution of, 12
  physiology of
    blood, 137, 138, 139
    brain, 150

circulatory system, 136
digestive system, 142,
  143
ear, hearing, 153, 154
eye, vision, 151, 152
heart, 135
hormones, 156
kidneys (excretory
  system), 145
liver, 128, 142, 143
muscular system, 131
nervous system, 148, 149
pancreas, 144
reproductive system,
  157, 158, 159
respiratory system, 133,
  134
skeletal system, 129, 130
skin, 146, 147
special senses, 155
spleen, 144
teeth, 132
races of, 160
Manatee, 116
Mandrill, 110
Mango, 51
Mangrove, 44
Manioc, 50
Manta ray, **84**
Mantle, of molluscs, 64, 65
Maple tree, 35
Marigold, marsh, 33
Marijuana, 54
Marine mammals, 116
Marlin, **126**
Marmosets, 110
Marrow, bone, 129, 137, 138
Marsilea, fern, 27
Marsupials, 104, 106
Marten, 118
Master gland, pituitary, 156
Mastication, of food, 142
Mating, *see* reproduction
Mating flight, queen bee, 77
Mayfly, 79
Meadow saffron (autumn
  crocus), 36
Mechanical energy, 41
Mediterranean man, 160

— (stone) pine, **28**
Medulla, of brain, 150
  of kidney, **145**
Medusas, 61
Megatherium, fossil sloth, 10
Meiosis, 6
Melanin, 147
Mendel, Gregor, 14
Menstruation, 157, 159
Mermaid's purse, 84
Mesophyll, of leaf, 39
Mesozoic era, 91, 95
Metabolism, *see* photo-
  synthesis, respiration
Metamorphosis
  in amphibians, 89, 90
  in butterflies, 74
  in insects, 73
  in moths, 74
  incomplete, 79
Mice, 112
Middle lamella, of plants, 4
Midges, 78
Migration
  in birds, 103
  in butterflies, 74
  in fishes, 86
  in lemmings, 112
Milk teeth, 131
Millet (sorghum), 37, 49
Mineral salts, 125
Minerals, 140
Mink, 118
Mint, 34
Miohippus, horse, 11
Mistletoe, 42
Mites, 72
Mitochondria, 3
Mitosis, 6
Moa, 98
Moisture loss from body,
  146
  see also transpiration
Molar teeth, 132
Mole, 108
Mollusca, phylum, *see*
  bivalves, Cephalopoda,
  gastropods
Monarch butterfly, 74

Illustrations, acknowledgments and picture credits

Ardea: **10r** centre, **64r** right, **88v** right, **92v** top, **93r** below, **97v** top & centre. **r** below, **98v** right, **99r**, **100r**, **102v**. & **r** left, **105r**, **106r** top, **107v**. & **r** top, **108r** right, **109v** left, **111r** below, **114v** left. & **r**, **117**, **118r**, **121v** top. & **r** below, **122** left

Bruce Coleman Ltd.: **9v**, **13r** top, **15v**, **24r** right, **26r** left, **53r**, **56r** right, **58r** right, **66r**, **67v** top. & **r** below, **68r** right, **69r** right, **71r**, **72v** right, **73v** below, **75v** below. & **r**, **78v** top & centre. **r** centre & below, **79r**, **81**, **84r**, **89v**. & **r** right, **92v** below, **93v** centre & below, **97v** below, **102r** right, **106r** below, **110v**, **111r** top, **112**, **113**, **115v** below. **r** below, **118v** below, **119**, **120v** right. & **r**, **121v** below. & **r** top, **122** left, **125**

Gene Cox: **4r**, **19v**, **20r**, **22v** right, **42** below, **58r** left, **59v** right, **62v** left, **137r**, **149v**, **151r**

Diagram: **2**, **10v**, **16**, **40**, **41**, **57**, **103v**, **104**, **123**, **124**, **126**, **140v**, **143**

Barry Driscoll: **11**, **65**, **74**, **83**, **86**, **90**, **116**

Hawkely Studios: **5v** *(Kings College, London)*, **9r**, **13v** top, **26r** right, **27r** left *(Kew Gardens Library)*, **28r** top & below, **34r** right. & **36v** *(Kew Gardens Library)*, **44v** left *(The Wellcome Institute)*, **50v**, **54r** left *(Kew Gardens Library)*, **56v** *(Victoria & Albert Museum)*

Bridget Heal: **50r**, **72r**, **140r**, **141**

NHPA: **13v** below, **21v**, **22r**, **24v**. & **r** left, **28v** below, **32v** left, **35v** left, **37v** left, **38r** left, **44v** right, **46v** left, **51r**, **54r** right, **61v** top. & **r** top, **64v**. & **r** right, **66v** below, **71v**, **72v** left, **73r**, **76**, **77**, **78v** below. & **r** top, **79v**, **80v**, **84v** right, **85**, **87r**, **89r** left, **92r** centre, **97r** top, **98v** left. & **r**, **103r** right, **105v**, **106v** top, **107r** below, **108v**. & **r** left, **118v** top

Picturepoint: **13r** below, **17r** right, **19r**, **22v** left, **25v** right, **27r** right, **28r** centre, **44** right, **48r**, **49r**, **52r**, **55r** right, **58v**, **59r**, **60v**. & **r**, **61v** below, **62v** right, **63**, **64r** left, **67v** below, **80r**, **84v** right, **87v**, **91r**, **94v** right, **106v** below, **120v** left, **127**, **147v**, **160v** left. & **r**

Ivan Polonin: **38r** right, **88v** left, **92r** top, **109v** right. & **r**, **110r**, **115r** top

Oleg Polonin: **29v**. & **r** left, **32r** left, **33v** below. & **r** top, **34r** left, **35r** left, **37v** right. & **r**, **38v** left. & **r** left

Sydney Woods: **3**, **4v**, **8v**, **14**, **17v** left, **18v**, **20v**, **30**, **31**, **39v**, **42**, **47**, **49v**, **59v** left, **60v** right, **62r**, **66v** top, **67r** top, **68v**. & **r** left, **73v** top, **77r** right, **96**, **99v**, **100**, **129**, **130**, **131r**, **132v**, **133**, **134v**, **135**, **136**, **137v** right, **142**, **145v**, **146v**, **148**, **149r**, **150**, **151v**. & **r** right, **152v** right, **153**, **155r**, **156**, **157**, **159**, **159**

5r *Suzanne Stevenson*

6 *Suzanne Stevenson (University College, London)*

7 *John Norris Wood*

8r *Natural History Museum*

10r top. & below *Uniphoto*

12 *Natural History Museum*

15r *The Wellcome Institute*

17v right. & **r** left *The Innes Institute*

18r *Luton and Dunstable Hospital*

21r *Ministry of Agriculture*

23 *Oxford Scientific Films Ltd.*

25v left. & **r** *Foord*

26v *Foord*

27v *Foord*

28v top *Sonia Halliday*

29r right *Clare Leinbach*

32v right *Foord.* **r** right *Marcel Sire*

33v top *Sonia Halliday.* **r** below *Foord*

34v *Marcel Sire*

35v right *Sonia Halliday.* **r** right *Foord*

nuch